Poison Darts

Further information can be found at www.poisondarts.net

First revised edition, 2004

Printed in United States of America

ISBN 0-9742938-1-4

For Lisa and Nina,
may they always find newts and toads.

"The most effective agent in the extinction of species is the pressure of other species" — Alfred R. Wallace

Preface

This work consists of two volumes. Volume I is fiction and contains a not so serious adventure story about a non-profit organization dedicated to protecting nature preserves around the world. In addition, a revolutionary new kind of contraceptive is developed that requires taking a pill or capsule to invoke temporary fertility instead of temporary infertility.

The story is used as a platform to convey potential solutions to the planet's plight and to show how and why they would work. The characters and their adventures are a fictional matrix in which real sets of solutions have been embedded. The high-tech wizardry mentioned throughout the story already exists, or has the immediate potential to exist, or was being developed as this book was being written.

Volume II is nonfiction and can be read independently from Volume I. It is a collection of essays on the subjects of nature, human nature, evolutionary biology, and the synthesis between them that has put in motion the sixth great extinction event in the history of our planet.

Not being much of a conformist, I have paid little attention to existing protocols for notes. Traditionally, notes were useful only to people with access to large libraries. By referencing websites, I have made the notes useful to anyone with access to the Internet.

One downside to referencing websites is that many of them will disappear shortly after or even before this book is published. I have remedied this problem by providing continuously updated footnotes on the www.poisondarts.net website. You can put your book down, go to www.poisondarts.net, and click on the note that piqued your curiosity. You can, within a few seconds, be reading about the potential for spider webs to enhance body armor or go to the United Nations website to verify a population statistic.

ACKNOWLEDGMENTS

I would like to thank—not in any particular order—those who helped me with this book.

Leading the list is my wife Jane Lester. Jane did the preliminary grammar check that made the book presentable enough for others to review.

Ben Sayre would sometimes read the New York Times between pick-up games at a basketball court we both frequented. That is how we met. He turned out to be a graduate student in journalism. He did an absolutely excellent job of critiquing Volume I—what ideas he did not kill he only made stronger.

Dee Finley is an experienced writer and advised me throughout the project. She also happens to be my mom.

Mark Lorge, a sharp guy who is keenly interested in environmental issues and who also likes to dabble in cosmology, provided a great deal of intelligent critique. He especially liked the concept of a Sentinel suit and was responsible for major changes in the book.

Ed Glaze, a population activist and a walking encyclopedia of population issues, gave me much valued feedback. Glaze calls for extreme measures to control the human population. Although I acknowledge Ed for his expertise and thank him for his assistance, I do not—for many well defended reasons—agree with his premise that we should be promoting severe measures to deal with the problem.

David McGranaghan passed the book through the mind of a highly talented artist and provided insights no one else could. The incredible cave paintings in the south of France are evidence that artistic talent has been with us for a long, long time. Check out his site at NaturalistArt.com.

Jeremy Harris critiqued the book with a thoroughness that was just stunning. One of the smartest people I have ever had the pleasure of communicating with, his mind is like a steel trap; *nothing* escapes its attention. I was very impressed. I would like to thank him once again for his generous assistance. You can find several of his book reviews on Amazon.com.

Karen Pitts is also a population activist and owner of www.overpopulation.org. She reviewed the book while on a trip to Africa. She mailed the review copy back to me with comments in the margins and when I opened it, a pressed African flower fell out. I thought it was interesting that she was visiting Ngorongoro crater and the Serengeti while reviewing chapters that were also set in African game parks. She noted, for example, that the only gun she saw in Africa was the one carried by the night ranger to protect them from buffalo and elephants. She brought to the book a much-needed woman's perspective on world population issues.

Dan Murphy edited the prologue. Dan is a professional who actually makes a living as a writer. How he came to edit the prologue is an interesting story and I'm sorry that I don't have room here to tell it to you. But let me leave you with a hint. Having his name in the credits is going to draw sharp criticism from certain elements in the environmental debate.

I want to thank my brother Pat Finley for providing me with most of the material found in the chapter on San Diego and for being my nature-exploring buddy since childhood.

The frog on the cover is actually a realistic looking rubber toy that once belonged to my oldest daughter. She bequeathed it to her little sister a few years ago who subsequently loaned it to me for the cover photo. Although not a perfect rendition of a Phyllobates Terribilis, I felt it was close enough.

Prologue I

There have been at least five great extinction events in the geologic past: the Ordovician, Devonian, Permian, Triassic, and Cretaceous. The last one occurred 65 million years ago, and was caused by an asteroid several miles in diameter that created the Chicxulub crater now hidden beneath the Gulf of Mexico on the Yucatan Peninsula. The causes of the other extinction events are unknown but light is dawning as scientists vie for status and glory in a never-ending competition to unravel the mysteries.

In his book *Gorgon*, Peter Ward presents evidence that the Permian extinction was caused by a lack of oxygen. The mammal-like reptiles that inhabited the Earth at that time—the gorgon for one—suffocated, the primary piece of evidence being the red rocks found in the strata associated with the end of this period. Rust on a piece of steel is the result of oxygen being removed from the air and combining with iron to form iron oxide, which has a red tint. Interestingly enough, the rocks of Mars are red for the same reason, proof that it once had an atmosphere with much more oxygen. What pulled the oxygen out of our atmosphere is unknown. It may have been related to the CO_2 being released into the Earth's atmosphere by volcanic activity in Siberia, which in turn may have caused a rise in global temperatures which might have triggered a release of methane—all speculation.

The secondary piece of evidence supporting the Permian "lack of oxygen" theory is the huge lung capacity of the predominant surviving species—the pig-like creatures called Lystrosaurus. Greg Retallack, an ancient soils specialist at the University of Oregon, has hypothesized that they evolved to live in burrows, where carbon dioxide levels are high and oxygen levels are low. Without competition from other species, they came out of their burrows and took over the land.[1]

The cause of the sixth great extinction event—the one you and I are presently witnessing—is also known.[2] As with the Permian extinction, this one is not happening overnight. It has been underway for much longer than most people realize. Many of the larger mammals and birds that roamed the Earth as recently as 11,000 years ago disappeared shortly after the arrival of humans in their part of the world: the Americas, Asia, Australia, Europe, and numerous islands. The cave paintings in the south of France graphically depict humans hunting a plethora of now extinct animals. The more wary, secretive, or smaller, species that have managed to survive our onslaught to date are finally succumbing as well. The biodiversity all across the planet is being destroyed in an ever growing and accelerating apocalypse caused by us.

As far as I'm concerned, further quantification of this carnage isn't necessary. That job has been done and done extremely well by others. Those who have read these works are fully aware of what is happening. The rest of humanity is along for the ride. It is not my goal to convince anyone that the destruction of the Earth's biodiversity should be a concern. *This book was written for those of you who are already convinced and want to see solutions.*

I recommend that you read *The Future of Life* by E.O. Wilson if you are looking for elegant arguments for the preservation of our biodiversity. He did a far better job of it than I could ever hope to do.

Way back in 1991, eight people and thousands of species of plants and animals were sealed inside the giant terrarium called Biosphere II—the name Biosphere I being reserved for planet Earth.[3] The hypothesis was that the sealed ecosystem would reach equilibrium and provide the eight volunteers with all of

---

[1] http://www.newscientist.com/hottopics/climate/climate.jsp?id=ns99994138

[2] http://www.nationalgeographic.com/ngm/9902/fngm/

[3] http://www.accessexcellence.org/LC/ST/st4bg.html and http://www.library.arizona.edu/images/eng102/biosphere2/biosphere.htm

their Earthly needs—food, water, and air—indefinitely. Something quickly went wrong and oxygen had to be added. The adventure was called off two years later after the ecosystem broke down and could no longer sustain human life—the participants were essentially suffocating. The experiment was a failure… or was it? Had the hypothesis been that a *simplified* ecosystem could *not* remain in equilibrium and thus sustain human life, the experiment might have been hailed as a roaring success. It would be held up today as evidence that we need to maintain the *complexity* and *diversity* of Biosphere I or risk having it tip out of balance the way Biosphere II did. Ironically, the structures that comprised Biosphere II will soon be used to study the effects of global warming. Rising temperatures combined with an ever-growing list of extinctions are ominous signs that Biosphere I might be going out of balance.

I strongly suspect that humanity will find ways to house and feed itself for many generations to come. If this blasphemous statement relieves you of fears that your grandchildren may go hungry because of ecological collapse, you are now absolved from further concern about the planet's biodiversity. You should put this book down now. Your time will be better spent fulfilling the genetically predestined behaviors befitting our species: competition, status seeking, and reproduction.

If, on the other hand, you want to preserve the planet's biodiversity as insurance against future calamity, or if you believe that wildlife must be allowed to exist regardless of human wants and desires, then read on. I will be presenting ideas for preserving what is left, but few reasons for doing so. As I said earlier, others have written books suggesting why we must preserve it. I see no reason to repeat their arguments. Missing from all of these works are viable schemes for accomplishing this preservation. This book will help fill that void.

The solutions proposed here depend on the acknowledgment of the existence of human nature. Empathy for strangers and other species is part of the solution, but we also have a tendency to form into self-righteous bands—liberals against conservatives, Moslems against Christians. We want to control the timing of childbearing to provide the best possible environment for our children, but few of us are going to inconvenience ourselves to protect the planet's biodiversity.

Being asked to voluntarily reduce conspicuous consumption is rather like being asked to use abstinence as your means of birth control. Conspicuous consumption is a physical manifestation of an undeniable human urge. Take for example, two members of our intelligentsia deeply involved in the biodiversity crisis: Paul Ehrlich and E.O. Wilson. Being humans, they have high status needs like the rest of us. They both spend most of their waking hours looking to have those needs met, not through conspicuous consumption but through writing books and speaking before envious peers and adoring fans, dropping the names of the Nobel Prize winners with whom they associate. But their speaking engagements burn up prodigious amounts of aviation fuel and their books consume enormous amounts of paper. The manufacture of which consumes a great deal of water, wood, and energy.

Whether they have consciously acknowledged it or not, they are getting their needs met from their celebrity status. This is an acceptable way to obtain status in their peer group (the world of academia). In other peer groups, whether professional athletes or architects, status is obtained in different ways. It is as futile to tell people that they should limit their consumption as it is to tell them they should limit their sex. We really must be a little more imaginative than that.

Contraceptive technology has allowed us to continue to meet our sexual needs without creating a baby with every act. But how do you allow people to satiate their status-seeking urges (which are closely related to sexual urges) without allowing them to advertise their prowess with trophies—books, published papers, houses, remodels, or cars? The answer is revealing—you can't. Telling people that they must stop having sex and seeking status is no different than telling them that they must stop walking upright. You couldn't call yourself a normal healthy human if you stopped doing all of those things. You cannot change human nature.

It is equally futile to ask our politicians and government bureaucracies to *force* us to limit our consumption. That idea has been tried many times—communism being the most prominent recent example—and only a fool would advocate continuing such a course. The answer is to find ways for people to compete for status that minimizes the impact on the ecosystem and biodiversity.

In his book *Plan B*, Lester Brown gives an example of how this can be done—tax redistribution. Tax the hell out of gasoline while reducing other taxes so that the total tax burden does not go up. People will demand better gas mileage and the free market will quickly supply it. People will continue to strive for status but at least they will be driving *hybrid* SUVs. Air pollution will decrease, global warming will be somewhat ameliorated, and we will buy some time to develop alternative energy sources to replace fossil fuels. That is so much more creative than harping that we must all stop consuming so much.

As a side-note, my 9-year-old daughter just brought me an ad she found in a newsmagazine—she often watches over my shoulder as I write. Picture a stand of trees with a caption in the middle that reads:

FINALLY, A VEHICLE THAT CAN TAKE YOU TO THE VERY PLACES YOU'RE HELPING TO PRESERVE

It is an ad for the Ford Escape, the first hybrid SUV—an oxymoron if ever I heard one, but hey, it's a start.

People do not limit the size of their families to save the planet; they do it to suit themselves. Likewise, people will not alter their lifestyles to end extinctions. Once set in motion the ideas proposed in this book would be self-propagating, because they give people what they want while preserving the planet's biodiversity as a *side effect*.

Preserving biodiversity is *not* a simple trade off between saving wildlife or feeding people. It is a matter of money and power. Most people would agree that it would be better to provide humanitarian food aid to a famine-stricken African nation than to go into the country's game parks and slaughter its wildlife to feed the hungry.

There are many ways to deal with the problem of an elephant that raids a subsistence farmer's field. You can fence off the field, you can relocate the farmer and provide him with another livelihood, or you can shoot the elephant. The cheapest method would be to shoot the elephant—money. You could simply forbid the farmer to shoot the elephant under threat of force and thus motivate him to find a way on his own to make a living—power. If that farmer fails to find a solution to his predicament and goes hungry, was it because he wasn't allowed to shoot the elephant or was it because he wasn't given the means to build a fence or find another livelihood? You get to pick, but you cannot have it both ways. You invest in the farmer to prevent him from going hungry, or you shoot the elephant.

It is popular today to pretend that you can have it both ways; you don't have to invest in the farmer or shoot the elephant. You can simply *teach* indigenous people to combine conservation and family planning. All you have to do is properly inform them of the importance of sustainability and they will obtain equilibrium with their natural environment. In other words, you can resurrect the myth of the noble savage.

Unfortunately, that is not how people are hardwired. Turn people loose in a forest, remove natural limits to their population growth, and they will consume that forest. It is human nature to alter the environment. It's the same thing beavers do when they cut down surrounding trees to build a dam and flood a meadow. Picture what the world would look like with six billion beavers on it. With so many of us humans altering the environment, there is simply little room left for the planet's biodiversity.

In addition, when you lift people out of poverty you send them on an insatiable quest for more status—bigger homes, more material possessions, or possibly opulent leisure in the form of an exotic African safari. China provides an example of what happens when people are lifted out of poverty. Simply because they can, many Chinese now strive to own an SUV; a bicycle is no longer sufficient. In addition, their meat consumption has increased fourfold since 1990—a signpost of a country emerging from its labor-intensive agrarian history.

This suggests another innovative idea to channel human nature in a direction that could help preserve biodiversity. If the Chinese government were to realize the tremendous prestige and status to be attained by declaring that only hybrid cars could be sold in their country, they could kill several birds with one law.

China is the only country on Earth that could get away with something like that. Such a move would be deeply envied by Western nations, thus providing the Chinese with great status. Overnight, China would take the lead in innovation—effectively kicking our collective butts. Their huge market would greatly accelerate hybrid car technology and bring the price of these cars down dramatically. In an attempt to reclaim their prestige, Japan and Western nations would try to outdo them in hybrid technology. Instead of a pointless space or arms race, we would have a hybrid technology race—good stuff. China's air pollution problems would be on their way toward resolution and their energy consumption greatly reduced—all side effects from an instinctive pursuit of status.

Once again, an idea like this is much more appealing than continuing to rail at our governments to make us stop consuming. Technology and innovation nurtured in an environment that acknowledges human nature for what it is can accomplish great things.

Let's compare the life of that subsistence farmer I mentioned earlier to my own. The farmer, his wife, and two children farm ten acres on the edge of a jungle preserve. Their acreage has been cleared and will not be allowed to grow back. Regardless of what sustainable-living enthusiasts want to believe, unless forcibly prevented from doing so, the farmer will eradicate any animal, be it mammal, bird, snake, or insect, that he finds in his fields. He will also hunt to supplement his diet, killing monkeys and whatever else he can to provide protein and variety to his diet. Although he lives in a simple thatched roof hut he has to forage for firewood, and his cooking fires generate significant air pollution, just like any wood stove, only far worse. He will clear the land using slash-and-burn technology and will occasionally ignite large, air-polluting fires to eliminate slash from his farmland. His consumption of wood is continuous. It does not end once he has built his house.

In contrast, my family eats food grown on industrial farms that are so efficient they have outpaced the population explosion. We obtain protein from domesticated livestock, which is many times more efficient than hunting wildlife for protein. If we all hunted for our protein like the subsistence farmer, there would be nothing left alive on this planet bigger than a grasshopper.

We live in a timber-framed building where the lumber came from second- or third-generation trees that are replaced by new seedlings after being harvested. We understand that preventing all forest fires can have serious ramifications and support properly managed timber harvests as being preferable to a forest fire. We cook and heat with clean-burning natural gas, and I work out of my own home office. Not only is the standard of living in my neighborhood much higher than on the farm, it also has a population density that is ten times higher.

Both the subsistence farmer and my own family have usurped part of the planet's ecosystem to feed and house our families; it is only a matter of efficiency. Who is doing the most damage to the planet's biodiversity, the subsistence farmer or me? I don't honestly know the answer to that, given that my standard of living is so much higher than his, but, if that subsistence farmer and his family could move into a two-bedroom condo and be gainfully employed in a city, his damage to the environment would be diminished. The world could not be fed using subsistence farming. That much I know for sure. Urbanization can be good for the planet. The problem is that there are not enough *worthwhile* jobs. Thus, about a third of humanity still ekes out a living from farming.

It is currently popular to place most of the blame for environmental woes on the United States because we are the most consumptive nation. I do not see this as a productive path to preserving biodiversity. *Blaming* people, weather they are rich or poor, will not get us anywhere. People are the same the world over. When a San bushman of the African Kalahari was asked why he hunts, his response was not what you might expect. He hunts to impress the women back at camp with the meat he brings home, hoping to parlay his status as a skilled hunter into sexual liaisons or multiple wives.[4]

A quarter of a century ago I arrived in Seattle with big plans to live the most ecologically sound life possible. I started by renting a small trailer—8 feet by 25 feet—a mile or so from my

---

[4] See Human Evolutionary Psychology, by Louise Barrett, Robin Dunbar, and John Lycett

job. I rode my bike everywhere, rain or shine. My small car would sit idle for weeks or even months at a time. I went to great lengths to recycle everything I could. I bought used kitchen utensils from a junk man in my neighborhood. I also did not intend to have children. After four years of this lifestyle, the futility of it all slowly dawned. Potential mates were not particularly impressed and people were not lining up to emulate me. Idealistic young environmentalists did not populate the trailer park; low status losers did. Something was not right here. I could not see how my living like this was going to prevent the next extinction.

I now have two children, live in a house, and drive an SUV. The SUVs of today, by the way, weigh less and have smaller, more fuel-efficient engines than the station wagons my parents drove—they also happen to have air pollution controls and seat belts. Environmentalism is not a religion, and I am not a fallen angel. I have simply come to some defensible conclusions. I walked the talk, but the sacrifices were quite obviously not worth the effort to me as an individual and my experience can be extended to others.

The following example sums up the real problem. I drink two or three cups of coffee a day. The Internet is rife with the controversies stirred by my cups of coffee. Was the coffee "shade" grown?[5] Were the farmers paid a fair price?[6] Those two cups when multiplied by 3 or 4 billion other people drinking coffee have destroyed millions of acres of rain forest and disrupted the migratory patterns of some bird species. The underlying problem is not so much how we live; it is that there are so many of us.

---

[5] http://www.seattleaudubon.org/shadecoffee/
[6] http://www.oxfamamerica.org/campaigncoffee/art3391.html

The problem

How do you lift three billion humans out of abject poverty, and at the same time preserve what is left of our biodiversity? The fact that our population is expected to increase by an additional three or four billion people by 2050 is also problematic.

The solutions

Preserving biodiversity:

1) Greatly accelerate the acquisition of critical ecosystems.
2) Get existing preserves out of the hands of governments and into the hands of *non-governmental*, *non-profit* organizations—NGOs—that are legal entities inside the countries containing the preserves. The profit motive must be uncoupled from ownership to have any hope of saving a preserve.
3) Create an international security force to patrol and protect the preserves—analogous to UN peacekeeping forces but based on private funding and staffed with qualified and possibly deputized volunteers looking for adventure and a self-righteous cause (harnessing human nature). In other words, protect what is left of biodiversity with private property laws and a leviathan.[7]
4) Hold these preserves with religious fervor (again harnessing human nature) against the raging of human societies for the hundreds of years that it will take for global population to peak and then decline to a sustainable level. NGOs do not have a limited lifespan the way wealthy landowners and their private preserves have always had. Let humanity fight it out in the areas between preserves, but not in them.
5) Accept that the "noble savage" is a myth. Statistically speaking, humanity will not make personal sacrifices for unrelated people elsewhere on the planet and certainly

---

[7] http://en.wikipedia.org/wiki/Thomas_Hobbes (a benevolent authority to keep our hand out of the cookie jar)

not for other species. We have yet to prove that we can live in harmony with the planet.

6) Continue to create and use technology in ways that accomplish biodiversity preservation by allowing human beings to meet their needs with less strain on the planet—getting more for less. For example, telecommuting via the Internet has the potential to save billions of gallons of oil and get millions of cars off of the streets.
7) Lose the mindset that our ecological problems are the result of evil corporations, capitalistic societies, or non-vegan diets. Our problems are caused by people trying to profit at the expense of others, regardless of the social institution. The problem is human nature and many billions of people competing with each other. The ecosystems of China, India and the former Soviet Union are disasters that make Western countries look like a paradise. You cannot blame their environmental problems on capitalism or corporations, since the damage was done under communism in China and the USSR and that India is still primarily an agrarian, vegetarian society.

Reducing poverty:

1) Greatly improve our contraceptive arsenal. Contrary to popular belief, we need much better technology in this area.
2) Accelerate existing efforts to educate poor women and provide family planning.
3) Promote urbanization by improving infrastructure in cities and providing meaningful jobs. People are already leaving the land and flocking to cities. Promote that process with better cities. We should live in cities and visit the country, not the other way around. Subsistence farming is not some idyllic lifestyle, but a difficult, unpredictable way to make a living.
4) Find ways to distribute wealth that mesh with human programming. Giving food away promotes dependency and creates a welfare state. It is not a long-term answer.

> Creating decent paying jobs, whatever that takes, without chewing up more of the planet, is a better solution.

Of course, there will be unending attempts to gain access to the resources inside the preserves. Balancing that will be those willing to protect them—human nature in equilibrium. The only difference between this scenario and past human history will be the introduction of contraceptive technology with the capacity to break the age-old cycle of local overpopulation and its accompanying ecological destruction and warfare.[8] This is truly our last chance; extinction is forever.

Obtaining consensus is not possible. Others will contest everything I have said here. It is human nature to disagree, fight, argue, and compete. We have no other option. We cannot stop deluding ourselves that we are always right and everybody else is wrong. We splinter and disagree primarily because it is our nature to do so. Let us go on deluding ourselves, but let us do it outside protected parts of our planet.

Those who are uncomfortable with the idea of using a leviathan to protect nature preserves should heed the lesson learned by a young Steven Pinker. In his book, *Blank Slate—The Modern Denial of Human Nature*, he describes what happened to a normally law abiding Canadian province the day the police force went on strike. The first bank was robbed the next morning followed by widespread looting and the closure of all banks.

Ironically, the late Julian Simon, a conservative economist who was not particularly concerned about biodiversity, may have inadvertently provided a key argument supporting conservationists in their struggle to save what is left of it. Part of Simon's philosophy was that human beings are creatures of unbounded creativity. In a free market system, entrepreneurs seeking profit will always compensate for shortages of things like copper, wood, or oil through new innovations. If Simon was right, then arguments *against* roping off chunks of our ecosystem because we need the resources inside them are moot. We do not need the resources found in the National Arctic Wildlife Reserve or old growth forests to maintain healthy economies.

---

[8] Constant Battles by Steven A. LeBlanc with Katherine E Register

This gives conservationists a much needed platform to protect what is left, allowing humanity to get on with the business of finding alternatives, including ways to feed more people without consuming more of the planet.

Unfortunately, profit motive always follows the path of least resistance. It is usually more profitable in the short term to cut down old growth forests than it is to make lumber out of recycled plastics. The pressure to exploit nature reserves will never end. Analogously, slavery was also the result of people attempting to make a profit at the expense of other entities. Laws against slavery finally forced profiteers to find alternatives.

Similarly, labor laws have stopped manufacturers from forcing people to work more than eight hours a day, 40 hours a week without additional compensation. But the pressure to work employees harder never ends, as witnessed by the growing number of hours the average American now works. Laws are all that prevent a free market system from evolving back into virtual slavery. If laws preventing slavery successfully blunted the profit motive, then laws preventing the exploitation of preserves could be similarly successful. Slavery could not end until enough people were convinced that it was wrong. It is time to move to another plane of morality and realize that destroying the biodiversity of a planet is not only wrong but dangerous.

One way to lessen this pressure—or alternatively make it more difficult—to consume our nature preserves is to put them into the hands of private non-profit, non-governmental organizations. People have an inherent tendency to support private property laws. Private ownership provides a layer of protection that government would not. Witness how close the Bush administration came to drilling for oil in the Arctic Wildlife Refuge in 2003—just to obtain a few years worth of petroleum.

The importance of family planning

Another way to lessen pressure on preserves is to decrease the demand for what is inside them, mainly by reducing population growth through family planning and poverty reduction. There is strong evidence that reducing poverty does not increase fertility when combined with family planning. An educational opportunity, combined with family planning, is the best way to empower women and allow them to escape poverty.

To put family planning into perspective, 76 million (76,000,000) people were added to the world in 2003 and there were also roughly 40 million (40,000,000) pregnancy termination procedures.[9] It is quite obvious that without the right and the means for women to time births, the world's population might already be closing in on nine billion, and there would be far fewer places left to preserve—if any.

Conservation, family planning, and poverty reduction are three legs of a pedestal supporting humanity's future. Our population has already surpassed six billion and is rapidly heading for seven. The latest data from the population division of the United Nations predicts a world population increase of about 50 percent in the next five or six decades—peaking somewhere between eight and ten billion.[10]

Few moments are more joyous than the birth of a wanted child. Inversely, few moments are more brutal than the birth of an unwanted child to an impoverished woman who already has malnourished children she cannot adequately care for. Worldwide, at least half of all pregnancies are unplanned. The exact definition of "unplanned" is debatable, but birthrates would plummet if there were a way to make unplanned pregnancies a thing of the past. Try researching this statistic on the Internet. Some pro-life Web sites post statistics suggesting that 10 percent of pregnancies are unplanned. Other pro-choice sites claim that unplanned pregnancies may be as high as 70 percent. The actual number is likely somewhere in the middle and varies with time and culture.

Here in the United States, where every conceivable birth control method is readily available, the percentage of unplanned pregnancies is about 55 percent.[11] Most people are shocked when they see this statistic. Their first reaction is disbelief. But that statistic is accurate and certainly confirms the observation that

---

[9] http://www.agi-usa.org/pubs/journals/25s3099.html

[10] http://www.un.org/esa/population/publications/wpp2002/WPP2002-HIGHLIGHTSrev1.PDF (click on highlights and go to page 6)

[11] http://www.agi-usa.org/pubs/journals/3124699.html and http://www.popcouncil.org/publications/pdr/usfertility.html

there is plenty of room for improvement in our existing contraceptive arsenal.

You might call it a "take it and forget it" contraceptive.[12] One pill or capsule would impart protection for years, or even decades much the same way a vaccine works.

Unlike today's contraceptives, which temporarily impose infertility, this new type of contraceptive would make infertility the default mode, with fertility temporarily re-imposed when pregnancy was desired. For example, a contraceptive could deliver antibodies that would attach themselves to and deactivate sperm cell flagellum. Once taken, the contraceptive would require an anti-viral-type “antidote” pill to temporarily neutralize its effects thus reinstating fertility. In other words, people would have to take pills to initiate a pregnancy, rather than to avoid it.

Such a contraceptive must have the following characteristics:

1) Duality—a version for men and women.
2) Longevity—effective for very long periods.
3) Reversibility—at least temporarily.
4) Simplicity—in the form of a simple pill or capsule.

A contraceptive meeting these criteria would have the immediate effect of improving the quality of life for billions of people by allowing them to avoid unintended pregnancies, while giving those who want children the freedom to pursue their dreams. The availability of such a contraceptive would also test the common assumption that the high birth rates seen in developing countries are a simple matter of choice.

One reason this new contraceptive would be more effective than past technologies is because men will share the responsibility for birth control. The fact that a couple could both use contraception would greatly enhance its efficiency. In addition, a male version of this contraceptive would go a long way toward diffusing opposition to its development by women's reproductive rights groups.[13] The characteristic of reversibility would also help circumvent that potential roadblock.

---

[12] http://www.agi-usa.org/pubs/journals/3209300.html
[13] http://www.izew.uni-tuebingen.de/bme/96010108.htm

A few years ago I had to look long and hard to find any data on the topic of contraceptive vaccines. Since then, the Internet has exploded with information. They are primarily being used as a humane way to control a wide assortment of animal populations—both in the wild and in zoos. Research on the subject is growing rapidly. The last time I typed "contraceptive vaccine" into my favorite search engine I found thousands of references.

It cost $100 billion to put men on the moon.[14] It cost $20 billion to split the atom.[15] In comparison, the human genome was unraveled for less than $3 billion and the information posted on the Internet for all to use—an accomplishment that may someday prove to be far more valuable than either of the two previous examples.[16] However, a similarly significant U.S. government investment in developing a new and better contraceptive is highly unlikely, due to our polarized political environment.

The United States Agency for International Development (USAID) is an independent federal government agency and the only agency actively involved in population and reproductive health programs worldwide. It operates on less than one-half of one percent of the federal budget, compared with the nearly 50 percent that was dedicated to military expenditures in the last fiscal year.[17]

However, a new contraceptive would create tremendous profit potential for the company able to develop and market it. Private enterprise holds the key for its development.

We are presently living through an unprecedented population explosion, one that would have ended catastrophically long ago if we were not also living through an even greater technological explosion. I sit here typing on a computer that is connected to the rest of the planet via the Internet. Sitting on top of the computer is a wireless router sending the Internet through the air to my daughter's laptop as she flops on the couch to do her homework,

---

[14] http://www.asi.org/adb/m/02/07/apollo-cost.html

[15] http://www.brook.edu/dybdocroot/FP/PROJECTS/NUCWCOST/MANHATTN.HTM

[16] http://www.genoscope.cns.fr/externe/English/Actualites/Presse/140403_1.html

[17] http://www.warresisters.org/piechart.htm and http://www.usaid.gov/about_usaid/

completely untethered by wires or cables. I can store this book on a CD, ZIP cartridge, hard drive, or upload it to another computer hard drive. Sitting within arm's reach is a tiny cell phone, which also serves as a calculator, calendar and clock. Connected to the computer is the digital camcorder used to take the photos for this book's cover. It also has the capacity to record images in total darkness and a digital zoom feature that can greatly magnify objects with the touch of a button.

What made all of this technology available? Free enterprise. The Internet may have been launched by government research, but it was free enterprise that made it what it is today. Free enterprise is an extension of that aspect of human nature that goads us to continually compete for higher status and profit. These examples of technology are being matched by technology growth in the biomedical fields. We have eliminated smallpox, cloned seven species of mammals to date, and unraveled the human genome. Once we decide we want a better contraceptive, we can have it. It is my hope that this book will help stimulate interest and investment in the "take it and forget it" contraceptive and accelerate its development.

The timely introduction of this contraceptive would help insure that we do not exceed the median population estimate and would possibly facilitate the world's population peaking at the lowest estimate. It would also help to insure that our numbers do not start to climb again after hitting the peak.

Protecting critical habitat

The health of the planet requires that we allow the human population to shrink. The dream of eliminating poverty and hunger depend on having a world with fewer people living on it. But we must also protect what is left of the ecosystem until our population peaks, stabilizes and slowly declines to some as-yet undetermined number. This problem is being addressed by thousands of conservation organizations around the world—the largest and best known being The Nature Conservancy and Conservation International.

The Nature Conservancy controls 11.5 million acres. This sounds impressive until you realize that these holdings are broken up into thousands of isolated parcels spread out all over the United States and other parts of the world, the majority of them smaller than a pinhead when viewed on a standard size globe of the Earth. Most of them may be too small and too isolated to survive in perpetuity without major extinction events within their boundaries.[18] To put it simply, we must protect as much critical habitat from humanity as is humanly possible as fast as possible.

It has recently been estimated that for about $28 billion, enough critical habitat could be bought or leased to protect 70 percent of the known plant and animal species in the world.[19] In other words, for a fraction of the money our government just spent in Iraq we could save the planet's biodiversity for future generations. Will we let this opportunity slip away?

Even if we could gain control over enough habitat, owning the land, leasing it or declaring it a preserve does not automatically guarantee protection. A preserve must be actively, physically protected. Consider these recent situations:

- Indonesia is losing an area of rainforest the size of Switzerland every year. More than 75 percent of the logging in that country is done illegally, the government being too weak to stop it. Nearly every preserve in Indonesia is under assault, with some estimates claiming that there will be no harvestable trees left within eight years. Some parks have been illegally logged so badly that they are no longer viable as nature preserves.[20]

---

[18] The Song of the Dodo: Island Biogeography in an Age of Extinctions by David Quammen.

[19] http://news.nationalgeographic.com/news/2001/05/0503_eowilson.html

[20] Irony and justice? Fri Jan 2, 2004 Los Angeles Times By Richard C. Paddock Times Staff Writer http://www.mongabay.com/external/tigers_indo_2004.htm

- Old landmines and mortar rounds are being used to poach the last remaining tigers in Cambodia's preserves.[21]
- Edward Wilson tells a story in his book *The Future of Life* about a team of researchers who accidentally landed their helicopter in a camp of poachers who were also searching for the same *non-existent* Sumatran rhinos in a "wildlife-protected area."
- Thanks to new roads, bridges, and a gold rush, a preserve in Myanmar (formerly Burma) is now inundated with tens of thousands of people. Wildlife is being slaughtered for food and dynamite is being used to catch fish. Hunters and soldiers trip motion sensitive cameras—hidden in the jungle to count wildlife populations—more often than they are tripped by wildlife.
- Industrial-grade emeralds were recently discovered in Ankarana, a park in northern Madagascar. In less than one year over 12,000 miners had inundated the reserve. Elizabeth Royte, in her book *The Tapirs Morning Bath*, summed it up when she wrote, "… Ankarana was protected only on paper. When a value with more immediate payoff than biodiversity came along, the 'protections' evaporated and lemurs lost out."

A common refrain heard around the world is that there aren't enough resources to adequately protect game parks and nature preserves. We can start by improving on that situation, as these examples show:

- A full time staff of guardabosques—camouflaged and armed forest guards—protects Barro Colorado Island, a preserve in the middle of Gatun Lake in Panama. Without armed guards and the lake acting as a protective mote, there would be little left in this preserve to study.

---

[21] http://www.felidae.org/NEWS_ARCHIVE/FEER_CAMBODIA_--_June_21__2001.htm

- The elephant populations in Uganda's game preserves were once poached so heavily for their ivory that only elephants which grew small tusks or no tusks at all were left to pass on their genes—a textbook example of evolutionary selective pressure at work.[22] It took heavily armed anti-poaching patrols and hundreds of miles of fences to stop the slaughter. Ironically, we are presently experimenting with contraceptive vaccines to control the numbers of those elephants now trapped inside game parks for their own safety.[23]

Musing over the latest UN human population estimates, David Quammen, author of *Monster of God*, wrote, "Call me a pessimist, but when I look into our future, I don't see any lions, tigers, or bears."

---

[22] http://www.asa3.org/archive/evolution/199809/0534.html
[23] http://kenya.com/wildlife/wildlife_004.htm and http://www.ovpr.uga.edu/researchnews/spring98/elephant.html

Volume I

Contents

Volume I

Volume II

CHAPTER 1

Choco infertility

While performing my duties as expedition physician on a field trip to study the Choco Indians of Colombia, I stumbled upon some information that had the potential to change the course of human history.

I was watching my Choco friend Qui apply poison to the tips of blowgun darts. The toxin is made from the frog family Dendrobatidae. Commonly called poison dart frogs, they come in a variety of colors, black, yellow, green, red, orange, blue, or any combination of these. Early researchers used to test the toxicity of different species by touching their tongues to the frogs. Typically, the only symptom was a slight numbness in the mouth area. The practice was abruptly stopped with the discovery of Phyllobates Terribilis, the most toxic species in this family. This tiny yellow frog contains enough poison to kill a dozen people.

Some Indians gently rub their darts over the backs of these frogs and then release them unharmed—just the kind of behavior one would expect to see from a noble savage. Qui's technique, however, was to shove the dart down the frog's throat and roast it over a fire.[24] The resulting residue is deadly and can last for a year.

As we sat by the fire, Qui confided in me that anyone surviving an accidental prick from a poisoned dart would be rendered infertile. The only way to regain one's fertility is to swal-

---

[24] http://www.cmzoo.org/aompoisonarrowfrog.html and http://www.geocities.com/Heartland/Acres/6113/history.html

low live dart frog tadpoles—which apparently are non-toxic—a day or two before having sex.

Shortly after my return to Seattle, I searched the Internet for a retailer of poison dart frogs in the area. I found what I was looking for at a site called www.poisondarts.net. An entrepreneur by the name of Edward Osborne owned the business. I arranged to visit his establishment. His address led me to a rundown house in a rough neighborhood. With some trepidation, I knocked on the front door. A skinny black guy with big round glasses answered.

"Hi, I'm Sarann."

"Pleased to meet you Sarann, I'm Ed. Come on in. So, you're into poison darts."

"Well, no. I'm more interested in frogs," I said as I closed the door behind me.

He admonished me with a raised eyebrow. "I was not talking about poisonous hunting instruments. I was referring to the frog genera Endrobates, Epipedobates, Phyllobates, and Minyobates.[25] What do you want them for?"

"I plan to use them for a toxicity study."

"You may have a problem there," he said while pouring himself a cup of coffee.

"How's that?"

"For reasons not fully understood, captive frogs lose all of their toxicity."

My bubble was popped. I stood there feeling like an idiot. "I suppose it was a stupid idea in the first place," I said, trying to hide my disappointment.

"Don't look so glum. You're dealing with a professional frog breeder here. It has long been suspected that they get their poison from something they eat. I've been experimenting, trying to unravel this mystery. José, my main man in Colombia, has been sending me samples of jungle invertebrates packed inside… ah, another product that I buy from him, and I've been feeding them to my frogs. I have finally nailed down what makes them poisonous."

---

[25] Q. Jansson – 2000, http://www1.tip.nl/~t272198/

I hadn't realized how dedicated Ed was to his craft. Imagine having to put "Professional Frog Breeder" as your occupation on a loan application.

"That's amazing. May I ask what it is they're eating?"

"You can ask all you want but I won't be telling you."

"I see… fair enough, but how do you test for toxicity?"

"That's an old trick. I just touch my tongue to the back of a frog and wait to see if my lips get numb. Observe…"

I watched as he reached into a nearby tank, pulled a beautiful blue frog out, and lightly placed his tongue on its back.

"Nothing to it. I can already tell this one is tanked up," he said.

I was appalled, but convinced. "Sell me that one, a red one, an orange one, and one of those yellow ones. You don't do that taste test on the yellow ones, do you?"

"Wha woo woo wink wi wam? Wupid?"

I didn't think Ed was stupid. An infertile frog breeder with numb lips, maybe, but not stupid.

I set up a lab in my basement and performed a series of experiments. I lightly rubbed mouse food pellets on the backs of the frogs. I then fed these pellets to white lab mice, carefully recording the color of frog used. The first thing I learned was *not* to rub pellets on the bright yellow frog. The surviving mice became infertile, both males and females.

My next move was to try to reverse the infertility. For this, I needed tadpoles.

I found myself back at Ed's house.

"Tadpoles?" Ed said. "That's pretty unusual. These frogs are difficult to raise from tadpoles. Frog breeding is not for everyone. It takes meticulous attention to detail and a lot of patience. You think you're up to the challenge?"

I assured him that I was.

Fighting off my guilt, I made tadpole smoothies in my blender that same afternoon. I added this concoction to the mouse water bottles and waited a few weeks. It worked. Both sexes became fertile while drinking the tadpole water.

I then hired a patent attorney, which I soon discovered is synonymous with "used car salesman." Many thousands of dollars later I had a patent for a "mouse contraceptive and its antidote"—not exactly what I was looking for. The patent examiner had insisted on some kind of proof that the contraceptive worked on people before he would issue a patent for a human contraceptive. He had suggested that he might accept expert testimony.

I realized at that point that I would need a little help. I flew back to Colombia to retrieve my good friend Qui. Fortunately, thanks to his association with various Catholic missionaries, he spoke decent Pidgin English.

Qui had never left the jungle before. He was very excited to do so, knowing full well that it would lead to years of material for stories around the campfire. Living in the jungle, one never gets to see anything at a distance. Qui could not adjust to the idea that far-away objects looked small. I was unable to convince him that cows seen in the distance were not tiny versions of the ones he had seen up close.

I will never forget the look on the patent examiner's face as we walked into his office. Picture a crisp white polo shirt with the heavily pierced and tattooed head of a Choco warrior sticking out of it. Qui gave the examiner many lurid and sexually explicit accounts of his adventures with his wives, two of which had survived accidental pricks from poison tipped darts. Both wives had become infertile in the prime of their reproductive years and both conceived shortly after getting up the courage to swallow tadpoles as instructed by the local shaman.

Qui is a very convincing storyteller. It's what Choco Indians do for entertainment every night instead of watching TV. He iced the cake by producing a dog-eared Polaroid of these two wives, both buck-naked, holding fat babies, and smiling for all they were worth.

It took another year, but I finally received my patent and immediately wrote letters to several pharmaceutical companies. The letter described my experiments with the poison dart frogs. I suggested that they investigate the possibility of developing a new type of contraceptive for people, one that provides lifetime protection. I also suggested that they develop in parallel an antidote for this contraceptive because it would have much less appeal without one.

For added incentive, I insinuated that they could give the contraceptive away, charging only for the antidote. I assumed that many corporations would jump at the idea of sterilizing people for free and then charging them for the cure. At first, the letters did not generate a lot of attention. The companies hadn't taken my word for it. They had been busy duplicating my results in their own labs. A few months later, however, I began to receive calls from corporate patent attorneys.

The potential profits from a contraceptive like this were astronomical. I picked a suitor—the S&B Corporation—and was flown to their lab in London.

My guide upon arrival was Benjamin Pinwright. Ben used to be a researcher, but was now director of the lab. He had all the social skills you would expect to see in a professional biomedical researcher. His attempts to emulate normal social behavior were admirable. My first name was at the beginning or end of every sentence he uttered, and his smile never wavered. He probably went home that night and packed his exhausted facial muscles in ice. A career in sales would have been a big mistake for him.

The lab had been studying dart frog toxins in an attempt to produce a painkiller more powerful than morphine.[26] Things hadn't been going well. Apparently, the frog toxin was proving to be a little too potent. The arrival of my letter had sent some of their researchers off into a new direction.

As we strolled down a long hallway, Ben explained how the dart frog contraceptive actually worked. "The amount of toxin needed for pain relief produces side effects that are… how do I say this… not trivial. The miniscule amount needed to produce an immune response however leaves the test subjects totally asymptomatic."

"Immune response? " I said, not yet grasping where Ben was heading.

"Yes, in other words, your body produces antibodies that attempt to neutralize the poison. What caught our attention was that these antibodies also attack sperm cells."

"Ah ha… sperm cells," I said as the realization dawned on me. "Could the fact that sperm cells look and act an awful lot

---

[26]http://psa-rising.com/medicalpike/frog.htm

like tiny tadpoles suggest that there is some distant evolutionary connection at work here?"

Ben, looking at me sideways with one eyebrow raised in derision, continued to elaborate. "Most Americans have received their vaccination shots. Today, oral vaccines have replaced many of the injections. Typically, these vaccines contain harmless, dead, or weakened viruses. Once they are introduced into our blood stream our bodies respond by generating antibodies against them thus making us immune to the dangerous virus. There are also vaccines to protect us from bacteria like Anthrax. Now, sperm cells share some common characteristics with bacteria. They even move about. A high fever will kill them just like it kills invading bacteria. That's what testicles are designed to do, you know, keep things nice and cool. That's why I prefer the loose feel of boxers to… "

I'd heard more than I wanted to at this point. "I think I'm getting the picture here. The idea is to use the poisonous secretions of the dart frog to develop an oral vaccine that would give long term immunity to sperm instead of immunity to a virus or bacteria."

"Yes, exactly."

"What about men?" I asked.

"Pardon?"

"Just a suggestion. I found that it sterilized both sexes."

"Well… maybe we'll consider that. Where was I? Oh yes. We have isolated a substance that can temporarily cleanse the blood stream of the antibody proteins that attack sperm cells. One can become temporarily fertile again by ingesting it."

"Let me guess. You found the antidote in the tadpoles," I said as we stopped briefly to get a cup of coffee.

"Yes, that is correct. The tadpole's blood is rich with an as yet unidentified chemical compound that suppresses the anti-sperm antibodies."

"Kind of an anti-anti-missile missile," I suggested.

"Excuse me?" Ben said, missing the joke entirely.

I continued, "What a coincidence. I mentioned tadpoles in the letter I sent to you."

"Letter?" He asked innocently as though he were reading an unfamiliar script. Ben was terrible at deception—yet another important social skill he had failed to develop.

"Never mind. Please continue," I said.

"Our trials are going exceptionally well. The vaccine we've developed has a perfect record on both mouse and primate test subjects. The only glitch we have encountered so far is your patent."

Still smiling, Ben continued. "It is my opinion that your patent is not applicable to our research because it failed to discern the underlying mechanism causing infertility."

"I see. What do your lawyers think?" I warily replied.

"They don't agree with me. They feel it is necessary to negotiate a settlement with you before we invest in further development."

"Is that why you flew me all the way out here?"

"Well, yes it is," Ben replied, completely unaware of my dripping sarcasm. "As a matter of fact we have a meeting scheduled for the end of this tour which is right about now...."

We'd just happened upon a meeting room, and Ben opened the door. "Meet our legal staff," he cheerfully announced.

I took a tentative sip of my coffee. Sitting on one side of a long table were four lawyers in nearly identical suits. I sat down, alone, on the other side of the table across from Ben.

The oldest looking of the four clones was the first to speak. "Let's cut to the chase. We are prepared to offer you a settlement of ten million dollars for exclusive rights to your patent."

The comment caught me with my coffee cup upended, mid swallow. Unable to breath or swallow, my reflexes took over. I was afraid this might happen. I was expecting they would offer me something pathetic, something I could laugh at.

While Ben wiped coffee spittle from his glasses, my mind groped for the appropriate response. Instead, I said, "Don't make me laugh."

The old clone was stunned. The stunned look faded into one of bitter resignation. He'd assumed wrongly that his adversary knew the score.

Before he caught his balance, I continued. "I could breed frogs and market them as 'Mother Nature's Natural Contraceptives.'"

He gave me a wry look and said, "Convincing people to lick

frogs and swallow tadpoles will take some doing, don't you think?"

"They could get used to it!" I shot back in defense. "As long as they don't lick the yellow ones," I added as an afterthought.

"Very well," the old clone drawled. "What terms would you like to propose?"

"My lawyers will be in touch," I said as I got up from the table. I guess I'd thought those words in daydreams so often that when the real opportunity happened along, they just came out of my mouth automatically.

I'd done my homework. If half of the people on the planet—about three billion—took this contraceptive in the first few years after its approval, it would generate 30 to 300 billion dollars in revenue depending on selling price, and that's just the tip of the iceberg. The antidote would be taken every day by couples wanting to conceive, for as long as that took, every time they wanted another child.

My net worth would increase significantly. With some funding in place, I would be ready for my next move, which was to form a think tank.

CHAPTER 2

Think tank

The settlement for my patent had included stock options as well as a percentage of profits from sales. The gamble had paid off handsomely. The value of S&B Corporation stock had gone ballistic in the year following my settlement. Suddenly, I was worth billions, and in a position to offer significant compensation to my tank members. They would need to be taken seriously. In our culture, there is no better way to bestow legitimacy than with lots and lots of money.

Not wanting my choices to be influenced by my own subconscious biases, I had done all preliminary interviews with job candidates via the telephone and the Internet. All resumes and qualifications had been thoroughly checked out. I did not actually meet them until their final interview. This would be my last chance to accept the candidates or reject them based on my intuition during a personal encounter. I wanted real thinkers. They could be fat thinkers, short thinkers, young, old, beautiful, ugly, black or white thinkers. I got what I asked for.

Dave

My first choice was an economist by training with a master's degree in economics from MIT. He was working for a government sponsored economic think tank at the time of our interview. On his resume, he had described himself as a square peg in a round hole with respect to his peers. This comment had caught my attention.

The door to my office opened and Dave walked in. I liked him on first sight. "You must be Dave Lohrax," I said while standing to shake his hand. "I'm Sarann Brown, pleased to meet you. Have a seat."

Dave was quiet, polite to a fault, and huge. He was about seven feet tall and had to weigh at least 350 pounds. I began by asking him for his thoughts on the role of large corporations in the destruction of the Earth's ecosystems.

He smiled and said, "This may not be the politically correct thing to say, as an environmentalist, I mean, but I believe that to get ourselves out of this mess, we must harness technology and the power of the free market. I believe that they are the only combination on Earth capable of saving it. Blaming our problems on capitalism strikes me as irrational. China and the former Soviet Union have created an ecological disaster that makes ours look tame by comparison. You certainly can't blame this on capitalism. Huge numbers of people and their natural wants and desires are the problem, not the institutions and economic systems we have created."

"Interesting perspective," I said.

"*Bite me*," he said.

"Pardon?"

"That's slang. It means... 'You don't say.'"

"Oh, I see."

"*Jackass*."

A moment of silence followed.

With his face in his hands, Dave had the look of a guy who had just blown the job interview of a lifetime. Without looking up, he said, "I failed to mention on my resume that I have Tourette Syndrome."

More silence.

I went to school for twelve years with a kid named Drew who had classic Tourette syndrome. He would cuss me out while flipping me off with both middle fingers as a matter of course. After about second grade, I didn't notice it anymore. Tourette Syndrome causes some of its sufferers to shout profanities and make obscene gestures uncontrollably and without provocation. Although the behavior looks like an angry outburst, it is just an unusual kind of tic, similar to a twitching facial muscle.

Switching subjects, I asked, "Married?"

"Are you kidding?" he said. "There's something about a guy who unpredictably screams obscenities that women find irresistible. Why would I get married when I can play the field?"

I wanted to laugh, but wasn't sure I should.

A grin slowly spread over Dave's face, giving me the go ahead.

"Can you tell me how you might contribute to this think tank?" I finally asked.

"Well, I have been researching some ideas to use tax exempt bonds to secure large tracts of ecosystem. Once the plan is cleared by the IRS, I believe we would have a powerful new economic weapon at our disposal."

"Sounds interesting," I said. "I have a joke for you. There was this guy who came across a shepherd and his flock. He says to the shepherd, 'I'll bet you $100 against one of your sheep that I can tell you the exact number in this flock.'

"The shepherd thinks it over and says, 'I'll take that bet.'

"'One hundred sheep,' says the man.

"The shepherd is astonished, 'OK, you win.'

"The man picks up the nearest animal and begins to walk away.

"'Wait!' cries the shepherd, 'let me have a chance to get even. Double or nothing that I can guess your occupation.'

"The man agrees.

"'You are an economist for a government think tank,' says the shepherd.

"'Amazing!' responds the man, 'but, how did you deduce that?'

Grinning, Dave interrupted me and said, "The shepherd says, 'Put down my dog and I'll tell you.'"

"You've heard it?"

"Many times. However, it has a logical flaw. The economist would have counted 101 sheep if he had truly mistaken the sheepdog for one of the flock and the shepherd would have told him to get lost."[27]

---

[27] http://netec.mcc.ac.uk/JokEc.html :Jeremy Harris, one of the editors for this book, pointed out that the economist would have counted 101 sheep if he had truly mistaken the sheep dog for a sheep.

"You have the job if you want it," I said. "The starting salary is five million a year."

The color drained from Dave's face. I was getting ready to do CPR when he finally spoke up.

"Lets do this."

Teresa

My second choice was Teresa Dawkins, a brilliant and somewhat controversial biologist from the University of Cambridge.

At the final interview, Teresa admonished me; "The planet doesn't need us to save it. Once the human race is extinct, new species will arise in a few million years from whatever is left of the flora and fauna; roaches, rats, pigs, chickens… whatever."

Nodding my head in agreement, I said. "The term, 'save the planet' is very misleading. Ever read the book 'The Mote in Gods Eye?'"

"No…" she replied carefully.

"You should. It presents a futuristic scenario similar to the one you just described. If you read it keep in mind that faster than light travel between the stars is just a fantasy."

She looked at me with a blank stare that said, "What-am-I-doing-here?"

I went on to explain what my actual goals were. "I want to stop the planet wide mass extinction of wildlife, and at the same time rein in human population growth, hopefully, in the next decade."

Teresa smirked. "Sure, we'll just start by banning all new development, logging, hunting, hiking and fishing clear across the planet. We will call ourselves the Planetoids and you can be Captain Planetoid."

I stifled a chuckle. I like this kind of sarcastic wit. Half hoping to elicit some more entertainment I went on to say that my ultimate goal is to improve the quality of life for all of humanity. She shot me a look of absolute disappointment, bordering on despair. From her perspective, I must have looked like a raving lunatic, which is probably closer to the truth than I'd like to admit.

Teresa continued. "Our population will peak in about fifty years. Even if our population starts to decrease at that point, it will take about a century for our numbers to shrink back to six billion. I have no doubts that there will not be an ecosystem left to save… "

As Teresa talked, my thoughts strayed. Thalidomide was first marketed in Europe in the late 1950's.[28] It was used to treat morning sickness during pregnancy until the discovery that it caused horrific birth defects. Although rare, thalidomide-induced birth defects still occur in nations where it is used to treat a variety of disorders including leprosy. None of my business really, but thalidomide use seemed the most likely explanation for why Teresa had no arms or legs. Adversity is sometimes the crucible of great intellect. It seems that Stephen Hawkings isn't the only wheelchair-bound intellectual to come from England recently.

Teresa was saying, "…there is very little hope that the biodiversity of the planet can be preserved. Many of the species that still cling to the planet are already doomed to extinction. They are essentially already extinct. There is no saving them. They are the last remnants of populations that are no longer genetically viable or populations that no longer have a habitat.

"Just picture what it will be like when there are nine billion of us. Picture the environment. We can only hope that our children and grandchildren will find themselves in a pocket of prosperity as the world population peaks. I do not hold out any hope for those billions of less fortunate people struggling to thrive or for the wildlife of our planet. I hope I'm wrong, but…."

I interrupted, "Do you want the job?"

"Is there any travel involved?' She asked.

"Not much," I said.

"I don't know. I'll have to think about it. What kind of compensation are we talking about?"

---

[28] Dark Remedy: The Impact of Thalidomide and Its Revival as a Vital Medicine by Trent D. Stephens, Rock Brynner.

"Five million for the first year—$208,333.00 every two weeks."

"I'm in. Where do I put my stuff?"

Jesús

The word theologian means different things to different people. The same can be said about the word atheist.

To some, an atheist is an evil psychopath, someone to be feared and loathed. To others, an atheist is simply a person who thinks that there is no deity manipulating the comings and goings of this bunch of upright walking primates that call themselves human beings.

To some, theologian is just a word that religious figures—pastors, priests, reverends, ministers, or rabbis—bestow upon themselves in an attempt to gain greater legitimacy. To others, theologian is a revered academic title, bestowed upon an individual who is interested in all things religious; one who studies religion with an unbiased point of reference. In reality, there are two kinds of theologians: those who study religion to confirm their belief that there is no God, and those who study religion to confirm their belief that there is a God.

My third choice was a theologian, a linguist fluent in seven languages and a world-class mathematician specializing in probability. He was also one buff looking dude. His long black hair and goatee reminded me of Zorro.

"What is your religious affiliation?" I asked.

"I was raised Catholic. However, I no longer identify myself with the Catholic faith."

"Why did you decide to leave the Church?"

"I had a revelation while doing my postgraduate research. It was this; Jesus Christ of Nazareth had to be one of three things:

1) He was truly the Son of God.
2) He was a con artist claiming to be the Son of God.
3) He was a first century version of a high functioning, schizophrenic street person who *thought* he was the Son of God.

"Which scenario do you think has the highest probability of being true?" I asked.

He continued without answering me, "Hobo's disease is described in medical literature as a mild form of schizophrenia. A high percentage of the hobos of the great depression were afflicted with it, and that is where the name came from. The same is true today for most street persons in America."

"Jesús, if the third case you described is the right one, there are a lot of people on this planet worshiping a wandering hobo. I take it you don't have a very high opinion of Jesus."

"On the contrary," he retorted, "history has always been shaped by personalities like Jesus of Nazareth. Just because he might be labeled with a mild mental illness today does not diminish his greatness. In all likelihood, many of history's greatest mathematicians, philosophers, and physicists were afflicted with a condition called Asperger Syndrome, which can be described as a mild form of autism.[29] One could argue that we are all crazy. It is just a matter of degree. Most of us just haven't been diagnosed."

"Really?" I said, "that's an interesting hypothesis."

Being mildly schizophrenic myself, I felt that he was hitting a little too close for comfort. My mental illness was diagnosed many years ago. I hear voices. Fortunately, they don't tell me to do things. I hear snippets of conversations, sometimes the cry of a child, or worse yet, quotes from Monty Python skits. What I find fascinating about them is their absolute crystal clarity. Sounds that clear can't be heard through the ear. Voices heard through one's ears always contain some background noise. The sound quality seems to suffer slightly being passed through the eardrum to the brain. I am constantly on guard for irrational thought patterns. I have yet to detect any. That doesn't mean I don't have them. I may be incapable of detecting them. I hope that the voices will remain my only symptom.

Finally, I asked what he thought his role in our great adventure might be.

---

[29] http://www.udel.edu/bkirby/asperger/aswhatisit.html

"Well," he sighed, "I would like to see my brand of religion popularized. The Moslem, Christian, and Jewish faiths were created long before we knew that the Earth circled the sun, that our sun is an ordinary star in a galaxy filled with billions of other stars, in a universe filled with billions of other galaxies. It is time to bring Christianity into the modern era, with a lot more emphasis on protecting this planet God gave us."

"OK," I said. "I guess we'll just have to see how things go. One last question. Does your name give you any clout in religious circles?"

"I would not be surprised," he replied. "I was named by my father who was drunk at the time. In my Mexican culture, Jesús is a common name."

"Well, yes," I said, ... "and your last name?"

"Right..." With some trepidation, he explained. "Having a last name of Christ is also not all that uncommon. Look in any phone book for a large city and you will find ten or twenty listings for people with the last name of Christ."

"Well Jesús, you're hired. Starting salary is five million."

"Sweet mother of God!" he said and then apologized.

There was no question about it; Jesús was an extraordinary individual. It was just my bad luck that he had such a potentially troublesome name.

I'm sure it was purely coincidental, but at this point my think tank was looking rather pathetic. I felt confident, however, that my choices had been sound. After all, I was not looking for normal people.

Bob

My last choice was the most controversial. His books on human nature, the environment, and the history of armed conflict were what caught my attention. A critic once described him as a "conservative, red necked, foul mouthed, homophobe." Granted, the description seemed to fit him like a glove. It wasn't in the name of tolerance and diversity that I decided to pick him. I needed him.

I perused his resume. "So, Mr. Smith, I see that you are a writer and a Boy Scout leader. You have a master's degree in mechanical engineering from...Purdue University. You were a lieutenant in the air force, speak Russian...and you are an active member of several conservation organizations."

"So far, so good," he said.

"Ever been married?"

"Three times."

"I see. I have heard you described as a homophobe. What's the deal?"

"Well, I'm not sure I deserve that particular label. The term homophobe suggests that I am afraid of homosexuals, but I most certainly am not. I have made a shit-load of enemies over the years, conservatives and liberals."

"I see. Have there ever been atheists or homosexuals in any of your Boy Scout troops?"

"Yep."

"Why?"

"Well, for starters, I am an atheist myself. As for the gay kids, I have admitted openly and publicly that I am uncomfortable with homosexuality. However, I believe that anyone so afflicted should be accepted as such. They have their feelings just like the rest of us."

I sneezed. "Excuse me, I have a cold. Another member of our think tank happens to be gay; can you deal with that?"

"Nary a problem."

If one didn't know better, one would swear that Bob was some yokel fresh out of hill country.

"Looks can be deceptive," he replied unexpectedly.

I'd heard about this other disturbing quality of Bob's—his ability to anticipate your thoughts. I was wondering at this point if Bob's back-woods demeanor might not be an elaborate deception.

"What did you do in the Air Force?" I asked.

"I flew A-10 wart hogs."

"And... what is a wart hog?"

"It's a slow, low flying jet designed to destroy enemy armor."

"Fun to fly?" I asked.

"There is no feeling in the world that can compare to the thrill of pulling your hog straight up, rolling inverted, and looking down on your target… unless it's the feeling you get when you squeeze the trigger that spews depleted Uranium munitions into that target."

Silence.

"Excuse me…."

"Yes, Bob?"

"Just what exactly is it y'all planning to do?"

"Ah… save the world," I said and then sneezed again.

"Would I get a paycheck?"

"Five million a year. Hand me that box of tissues."

"Five million what?"

"Five million dollars."

"You're shitting me, right?"

"No, I am not shitting you.

"OK then, where do we start?"

CHAPTER 3

First tank meeting

With the think tank fully staffed, we were ready for our first virtual session. I was in Seattle, Jesús was in Mexico City, Bob was on the East Coast somewhere, Teresa was in England, and Dave was in Wisconsin. Each member could see every other member in a separate window on his or her computer console. This proved to be a little awkward. We were not in the same room and this limited our ability to use subtle means of communication, like eye contact and body language.

I began, "Before we get started, I'd like to introduce Randy. Say hello Randy."

"Hello."

"He is our administrative assistant and computer wizard. He helped me find and hire you guys and now he is going to help run these meetings. You can't see him but he is listening in. His job is to make sure everything goes smoothly. If you have any technical difficulties, or need assistance of any kind, Randy is the man."

"So, what do you guys think about my birth control idea?" I said to start the ball rolling. "I'd like to call it the PDFC for 'Poison Dart Frog Contraceptive.'"

Silence.

Teresa spoke up first—this would be the start of a pattern. "Frankly, I think you should come up with a better name."

"Maybe you're right," I said, feeling slightly chastised, "any suggestions?"

"How about the '**T**ake **I**t and **F**orget **I**t' **C**ontraceptive?" Jesús suggested, "You could use TIFIC for the acronym."

"All right, from this point on I'll refer to it as the TIFIC."

"Excuse me. "

"Yes, Randy?"

"Could you give me a brief description of how the contraceptive works?"

"I'd be glad to. Think of it as an orally administered vaccination. It creates antibodies that deactivate sperm cells."

"How does it deactivate them?"

"It interferes with the operation of their little tails."

"I see. Can it be taken by men and women?"

"Absolutely, just like any vaccine. The antibodies in a woman's blood will deactivate any sperm that show up in her body."

"How long does it last?"

"Well, preliminary results suggest it may impart lifetime immunity."

"So, anyone taking it will be rendered permanently infertile?"

"I prefer to think of it as chronically infertile. You can temporarily reinstate fertility by taking a daily antidote pill until you conceive. Infertility will return a few days after you stop taking the antidote."

"I see. Is there anyway to regain normal fertility?"

"Yes, a complete blood transfusion."

"I got it now, thanks."

I continued. "If I may, I'd like to start the discussion by telling you a little bit about myself and some of my thoughts and concerns."

There didn't seem to be any disagreement to that so I went on. "I was a child when Earth Day and the word ecosystem made their first appearances. As a teenager, I decided I would live a simple, ecologically sound life as a vegetarian. I was going to do my part to preserve the Earth's ecosystems. The plan was to live in an Airstream trailer because they are made entirely of aluminum and no forest would have to be cut down to provide me a home.

"I graduated from Indiana University's School of Medicine on May 18, 1980. Immediately following the graduation ceremonies, I jumped into my rusted-out Pinto and headed for Seattle to start my Pediatrics residency at the University of Washington.

Unknown to me at the time, this was also the day that Mount St. Helens erupted. What are the odds that a kid from the flatlands of Indiana, who had never even seen a mountain before, would be forced to hole-up in Coeur d'Alene, Idaho because the highway was blocked by volcanic ash?"

"Roughly 250 million to one," Jesús interjected.

"Uh, right, thank you for that," I said, having been caught off guard. Every genius has his weaknesses. I would learn that for Jesús, it was recognizing rhetorical questions and idioms in the English language—one of seven languages that he spoke.

"I lived four years in an RV trailer and rode my bike in the Seattle rain to the bus stop to begin my daily commute. I would drive for miles to the nearest recycling center to donate my bottles and cans. I even stuffed the plastic products into bags and stored them under my trailer just in case they became recyclable someday. I was young, naive, and idealistic.

"One night, as the cars zoomed past me on my bike in the dark, pouring rain, I had a series of revelations. First, there was not enough aluminum ore on the planet to provide five billion of us—now there are over six billion of us—with Airstream trailers. Second, recycling is ultimately a pointless exercise because every baby born into this society will produce enough new waste and consume enough resources to nullify the life long recycling efforts of dozens of people. Finally, and I know this sounds rather pessimistic, people are motivated primarily by self-interest. People will not live in trailers and ride bicycles when there are other alternatives. Some people prefer to live in Yurts, or houses made of straw, sticks, or bricks. The answer does not lie in conservation, recycling, or vegetarianism. The answer is a smaller human population."

"Precisely," Teresa interrupted. "Not too long ago, Time magazine ran a cover story titled, 'How to save the Earth.' The article did not offer much in the way of real solutions, but their website had a survey that asked subscribers to pick from five items what they thought was the most critical threat to the environment:

Biodiversity Loss
Depleted & Polluted Water
Vanishing Forests
Pollution & Climate Change
Overpopulation

"Forty-four percent picked overpopulation. Now, I'm the first to admit that surveys of what people think often have little bearing on reality, but this survey had another fundamental flaw. Without overpopulation, the other items would be much less of a concern. The number of people on the planet was listed as just another threat to the environment when in reality it is the cause of the other threats. The continuing mass extinction and ecosystem annihilation would not exist if we could turn the population clocks back to 1945. At that time, the world had automobiles, airliners, antibiotics, vaccines, nuclear bombs, skyscrapers, television, and toaster ovens. There was technology aplenty and a booming U.S. economy. The items listed on that environmental survey existed, but not in quantities that threatened our children's and grandchildren's quality of life or the overall health of the planet. Why? Because there were only about two billion people on the entire planet, that's why."

I continued. "So, anyway, realizing the futility of it all, I settled into my mundane life as a pediatrician, commuting to work everyday, turning the pages on my day calendars to see the latest Dilbert and Far Side cartoons. In the back of my mind, though, I never stopped looking for solutions to the dual problems of ecosystem annihilation and billions of impoverished people.

"Dealing with teen pregnancies was a routine part of my practice. Studies consistently showed that about 55 percent of all pregnancies in the United States are unplanned. Virtually all teenage pregnancies are unplanned. This huge percentage of unplanned pregnancies in one of the wealthiest countries in the world is testimony to the gross inadequacy of existing contraceptive technology. The ideas of inserting bizarrely shaped objects into the reproductive tract (the IUD), surgically implanting a device under your skin every few years (Norplant), taking four shots annually (Depo-Provera), spermicides, diaphragms, condoms, and abortions will someday look primitive and brutish."[30]

I noticed at this point that Jesús appeared to be asleep. Had I been blabbering for too long? How do you wake up a sleeping participant in a virtual meeting? You can't inconspicuously lean over and nudge him with your elbow like you can in a real meeting. As a physician employed by a large HMO, I had learned to sleep through meetings with my eyes open, just like everybody else at the meeting. Apparently, Jesús had yet to learn that particular skill. I wondered if he had the sound turned off as well. I continued anyway.

"My experiences with some forestland that I own gave me the insight I needed to more fully understand the problem of humankind's destruction of the planet. We are just following our instincts to change and control our environment. We cannot change what we are, and that is why it will not be possible to halt the degradation of our natural ecosystems with this many people on the planet.

"When I purchased this property, I swore that I would preserve it. Nevertheless, over the years the urge to 'improve' this property has been difficult to resist. It now contains within its boundaries, an RV trailer, two cedar decks, and a hot tub. In short, I have made a mess of things.

"There are regulations to protect wetlands and control sewage. My neighbors paid not the least bit of attention to them. The bureaucrats in charge of enforcing the regulations did not agree with them and always looked the other way because they also owned property that they hoped to develop. I was having a hard time controlling my own urges to develop my land. How could I blame all the idiots around me as they built monstrosities on the edges of pristine trout lakes? What good was it to preserve my land if it was eventually going to be surrounded by development? I still have a recurring nightmare that I will walk over a hill at my property one day and find a shopping center there."

"*Asshole...* " Dave interjected.

I realized then that I had completely forgotten to mention Dave's Tourette Syndrome to the other tank members. Jesús

---

[30] http://www4.nas.edu/news.nsf/isbn/0309054427?OpenDocument

appeared to be wide-awake now, so at least he hadn't turned the sound off.

"I have to take a minute here to fix an omission I made on the personal information packet I sent to each of you. Dave has a condition called Tourette syndrome. Please ignore any profanities he may hurl in your direction during the course of this meeting. I strongly recommend that you look the syndrome up on the Internet after this meeting is over.

"My final point is this; bureaucracies, rules, and regulations are not going to save the planet. Running over foxes and building ugly cabins on trout lakes would not be so destructive if there weren't so many of us doing it. I'm done... any comments?"

No one volunteered.

"Teresa? Any thoughts?"

"Sure. As a British citizen, I have been watching U.S. population growth with great interest. I am a card-carrying member of NPG—Negative Population Growth. Initially, their goal was to reduce the world's population. They now deal primarily with U.S. population issues. Their studies had indicated that population growth in the United States was mainly due to immigration. They once sent me a newsletter which predicted that in just one decade the present rates of immigration were going to boost the population of the United States by the equivalent of about sixteen cities the size of Boston. I was appalled, but not nearly as appalled as when, ten years later, the 2000 census showed a population increase that was enough to fill *fifty* cities the size of Boston. The U.S. has just experienced the biggest population increase in its history. Thirty two million people were added since the 1990 census. I suspected that their lobbying efforts would fail for two reasons. First off, most politicians have now realized that cheap labor supplied by first generation immigration fuels a growing economy. Secondly, any pocket of prosperity like the United States will feel overwhelming pressure by those stuck on the outside of its borders to get in. I call this phenomenon 'prosperity osmosis.'"

"*Bitch....*"

"Uh... if the quality of life in Mexico were on a par with that in the U.S., the illegal immigration problems with Mexico would go away. The U.S. does not have a similar problem with the

Canadian border because they have quality of life parity. Of course, if Mexico had quality of life parity with the U.S., they would then have an immigration problem with Guatemala, parity with Guatemala would cause a problem with Panama, and so it would go.

"Cities and states experience a similar scaled down version of this osmotic pressure, or desire to move to a place where the quality of life is perceived to be better. Countries that think that humanity's burgeoning population is someone else's problem are badly mistaken."

"*Skank....*"

Dave's profanities were giving much needed punctuation to Teresa's long-winded comments, but a solution was needed here. "Dave, have you considered turning your mike off until you want to make a comment?"

"Will do," he said as he reached for the off button. "But I'd like to make a comment first."

"Go right ahead," I declared.

"Illegal immigration is a double-edged sword. Those same politicians who have learned that illegal immigration fuels a hot economy will also learn what a burden those tens of millions of unemployed people are during an economic recession. Illegal immigrants pay few taxes. All the same, more power to them I say. If I were stuck on the outside, I sure as hell would try to find a way in. These are hard working people striving for the same thing as the rest of us."

"Maybe someday, the world's borders will all be as free to cross as our state borders are," Teresa suggested.

"Fat... chance," Dave managed to say.

Bob had been conspicuously quiet. I thought I'd try to bring him into the discussion. Anyone who has read Bob's work knows that he possesses a powerful intellect. I feel that I know him through his writing. I hoped to someday meet the real man behind the mask.

"Bob, what are your thoughts on the TIFIC?"

"Well... if I'm right, and young men get the impression that taking this contraceptive will help to secure the attentions of women, then every young buck on the planet will find a way to take one. They'll do it if they have to steal it. Yes sir, a typical young man will do just about anything to increase his odds of getting into the pants of a receptive female."

"Amen to that." Jesús declared, drawing surprised looks from the other tank members. "Sorry."

"I don't get it," Randy said. "Why would taking this contraceptive give a guy an advantage over someone who didn't?"

"Jeez Randy," Lisa interjected. "It is all a matter of perception. If they *think* it might help, they'll take it. Ultimately it will be up to women to favor those men. Selection criteria by females can shape the behavior and even the look of males. The male peacock looks like he does purely because of female mate selection criteria."

"Uh… right," Bob continued. "Anyway, attempts will be made to limit access to the antidote in order to eradicate the spore of perceived enemies. I can almost visualize Israel withholding supplies of the antidote from Palestinians once this contraceptive has been introduced to them. We must have at least partial control over stockpiles of the antidote to counter that inevitable scenario through black market channels. We also need to get an operative into the pharmaceutical manufacturer's infrastructure as soon as possible, before they are done with trials and form a marketing strategy."

I was stunned, but at the same time, I was relieved. It looked like the real Bob was out. Apparently, he had heard enough to decide that this think tank had potential. I was flattered that he thought so.

"Do you think we also need to stockpile the contraceptive?" I asked.

"Not if the manufacturing data can be leaked to the Internet. I believe that the demand for it will be so great that black market channels will produce and distribute it to every corner of the planet regardless of any attempt to control it."

"Ok," I said. "I guess you get the first action memo. You on top of that Randy?"

"I am."

Bob addressed the group. "I'm confident that Sarann has not hired a bunch of fools. I need not stress the importance of secrecy on certain matters."

This would be my first taste of Bob's propensity for secrecy. Dave had disappeared from the monitor. "Dave?" He reappeared.

"Dave, you've been awfully quiet. Do you have anything to add before we move on to the next subject?"

He held a piece of paper up that read, "My microphone is inoperable at this time." That explained his apparent lack of enthusiasm.

"Randy, can you get that fixed?"

"Working on it."

"OK guys, I'd like to start another topic of discussion. Teresa, you told me at your interview that it would take hundreds of years to bring the population gently down to a few billion. You also suggested that the ecosystem would be irreparably damaged long before that point was reached. How did you conclude it would take so long?"

"It is pretty straight forward," she said. "Our population is not going to peak for another fifty years. At that time there may be nine or ten billion people. If half of those ten billion die from old age in the following fifty years, you still have five billion people left one hundred years from now. With a fertility rate of just one in that second fifty years, you will have added about two billion people so you would be down to only seven billion after a hundred years. You are not even close to getting to two billion yet."

"My God, I see what you mean," I said.

Teresa continued, "The theory of prosperity osmosis applies to those small pockets called wildlife preserves as well. Few will stand up to hundreds of years of humanity beating at their doors."

Suddenly, my telephone rang. It was Dave. "Very innovative," I said, as each of the other members picked up their telephones, "a five way conference call."

Dave started off; "I'd like to make one comment before we move on if I may."

"Go ahead," I said.

"Well, as an economist, I cannot predict *when* a stock market will take a dive, but I can predict with absolute certainty that it will. The same holds true for the human population. I just thought the analogy, you know, was a good one. I'll shut up now."

"No, that's great," I proclaimed. Dave was the least vocal member of the group. I didn't want him to feel overwhelmed by some of the more talkative members.

Teresa picked up where she had left off. "No one knows how big a wildlife preserve needs to be to protect its biodiversity and avoid extinction within its boundaries. The problem is too complex. Variables like weather, drought, and disease cannot be controlled or predicted. My own conclusion is that we must quite simply preserve as much as possible by interconnecting areas with wildlife corridors. Islands are vulnerable to extinction events because of their size and isolation. However, studies have shown that all wildlife preserves are essentially islands surrounded by a sea of humanity and they have similar extinction vulnerabilities. Isolated wildlife preserves may slow extinction events, but they cannot stop them. Animal population surveys in our very largest parks show that certain species that used to exist within the park boundaries are now absent. Extinction events happen too slowly to trip our instinctive alarm bells the way a burning house would, but a burning house is just what our planet is. The destruction is occurring at a pace that is below our time sense threshold."

"Could you elaborate a little more on this concept of time sense?" I asked.[31]

"Sure. We are incapable of seeing some events directly because they happen too slowly for our time sense to detect: plants growing, the movement of a clock's hour hand. It is no different really than our inability to detect sound frequencies that are too low for our sense of hearing to detect. Environmental degradation is happening too slowly for people to see, and therefore very few people are appropriately alarmed. Time sense is the defense mechanism of a sloth. It moves too slowly for many of its enemies to detect."

"Same technique I use." Dave said.

---

[31] http://hep.ucsb.edu/courses/ph6b_99/0111299sci-scaling.html and http://ourworld.compuserve.com/homepages/jmkenney/#PP

"Real cute," Teresa remarked, looking quite amused.

"These are all good points," I said. "The gradual development of suburbia is an example most of us can identify with. Many of us have memories of that wooded lot we played in as children that now has a house on it."

Jesús interjected, "Yes, to Mother Earth, human population growth feels more like a viral infection. HIV can be looked at as an antibody produced by the planet in an attempt to rid herself of a virulent infection called humanity. It may be a sign of the planet's growing desperation. We have moved from being one of the millions of symbiotic organisms sharing the same host to being a dangerous parasite. Time is running out."

"You sure come up with some interesting analogies," Bob said while lighting a cigar. "I think we're all preaching to the choir here. I have yet to hear any suggestions as to how we might protect what is left of the environment while the human population passes through its maximum in the next forty or so years."

"Good point," I said. "Why don't we start with you?"

"OK, first of all let's see what can be learned from other's failures. One scheme I am familiar with attempted to get farmers to preserve their piece of the Amazonian rainforest by collecting sap from rubber trees and transporting it to storage sheds along the river. The idea was that a boat would come down the river, buy their raw rubber, and everybody would make a nice profit, thus motivating them to preserve the rainforest. It failed miserably because the price of natural rubber was too low to bother with. A group of financial supporters visited one of these storage sheds to see how well the whole scheme was working and found to their horror that the shed contained jaguar skins instead of rubber."

Uncomfortable laughter filled the speakers.

"This isn't the least bit surprising to me," Dave said. "The natives found that the jaguar skins were far more valuable on the open market than rubber. We must find a way to use free market forces if we want any hope of success. I know of an example that has so far proven successful."

Suddenly, Dave dropped—literally—out of site off the computer screen. He re-appeared a few seconds later, hair slightly tousled. "Chair leg broke," he apologized and continued, "vast tracts of the Colombian rainforests have remained completely free of development, including all road building. The wildlife and ecosystem are stable. This wasn't planned. Free market forces inadvertently intervened. These forests are the strongholds of the Colombian drug cartel militias. They need these vast tracts of forest to hide in. There isn't a farmer, cattleman, lumberjack, hunter or fisherman in all of Colombia who will risk his life venturing into these places. The armed militias have inadvertently become very effective conservationists. They typically give trespassers one warning. Second time offenders are executed on the spot. The low supply and high demand for the coca crops that surround this rainforest provide the free market incentive that is protecting it."

I interrupted, "Are you suggesting that we get a piece of the action?"

"No, that's not my point," he said, mildly exasperated. "I am trying to say that we should use free market forces to fuel our schemes whenever possible."

"I have another model we might use with modifications," Bob said. "The Peace Corps. Admittedly, it isn't a conservation organization, but it has had a long life and has successfully marshaled a lot of resources. Their budget for this year exceeded 270 million."

"This may sound stupid, but what exactly does the Peace Corps do?" I asked.

Teresa volunteered. "Some see it as an adventurous lark for young adults from America's upper middle class. In many circles, having served in the Peace Corps is a sought after badge of honor. The volunteers are placed in an undeveloped country and work in some local health clinic or school or whatever. The Peace Corps provides them with travel expenses, housing, and food allowances. They are well tolerated by the locals and often appreciated as a free source of labor. Their presence also tends to boost the local economy. Judging by the number of books written about the Peace Corps by returned veterans, I would guess that a lot of the volunteers join to get material for a book."

"Hold on there Missy, they do some good," Bob suggested defensively.

"I guess that makes them a bunch of 'do-gooders.'" Teresa retorted. "My name, by the way, is Teresa, not *Missy...Bobby*."

I flinched. This was my first taste of the sparks that were going to fly between these two. "How do you see it as model we can use?" I asked Bob.

"Don't know yet. I haven't worked out the details."

"OK... why hasn't anyone brought up Conservation International, or the Nature Conservancy?" I asked. "What's wrong with using them as a model?"

"Hello?" Dave's voice suddenly boomed out.

"We can hear you," I said, growing a little weary of his shenanigans. He had fixed his microphone.

"We would have to be careful. All large organizations that are uncoupled from free market forces experience a degradation of efficiency: governments, large corporations, and non-profit groups. I have witnessed conservation organizations using helicopters—which rent for hundreds of dollars an hour—to airlift stepping stones to high altitude meadows just to make paths so that hikers would not damage the delicate alpine soil. It seems to me that a 'no trespassing sign' might have been a little more cost effective. I have also heard of feral pig hunts in Hawaii using rented helicopters. Can you imagine what it must cost to hunt pigs using helicopters? Not many of the local pig hunters can afford to do that. They usually *hike* into the forests to do their hunting."

Teresa waited a few seconds, not sure if Dave had finished. "Conservation organizations are in the business of protecting critical habitat. Many of their acquisitions, however, are small, isolated parcels that can be vulnerable to gradual extinction. We should use a different set of criteria when deciding what part of the planet to protect. For example, I believe we should go for bigger, less dramatic parcels that have a good chance of healing, fallow farmland on flood plains for example. We should look for areas that can be protected from encroachment and have the potential to link up with other protected parcels with road free corridors. Most of our remaining forests are on marginal land that people couldn't develop if they wanted to, typically mountainous areas, dry desert areas, or impenetrable swamps. Preserves created from fallow farm and flood plain land would soon be rich, diverse ecosystems again.

"Roads are ecology killers. Wherever roads are built, people in trucks with six-packs soon follow. Roads are pathways for infection. This is as true in the Amazon basin and the African Congo as it is here in the United States. We need to find ways to circumvent some county and state owned road systems that are isolating pockets of healthy ecosystem. Freeways are ribbons of death for fauna. They form barriers to migration and movement."

I was forced to interrupt. I saw no signs that Teresa was running out of wind. "We are just about out of time. Before we adjourn, I would like to ask Jesús for a brief description of his religious views and what his long-term goals might be. Jesús?"

"I will certainly try. My religion does not worship a white male primate of the species Homo sapiens, resplendent with robes and beard, residing somewhere in the sky. My God made all living creatures, not with a magic wand, but through the combination of unimaginable lengths of time and the raw mathematics of evolutionary selective pressure.

"Reverence for the life of the planet will be the centerpiece of my religion just as the belief that Christ was the Son of God is the centerpiece of other versions of Christianity. Jesus of Nazareth will continue to play a central role, not as the Son of God, but as the philosopher he was. The Bible was not written by Jesus, it was written by deeply superstitious, deeply ignorant, religious devotees. I'm going to bring it up to date. Following the Christian model, I fully expect to see groups splinter off, providing a version to suit everyone. For those who want to believe that their God tosses random miracles to those who successfully ingratiate themselves with prayer, so be it. I expect to see sects that believe in an after-life for those who need relief from the fear of death."

Bob blew smoke into the camera. "Why bother with another religion? We could put Sarann's birth control serum into tranquilizer darts and let rival religious groups fight each other with tranquilizer guns. They could render one another sterile with a dart in the ass. Northern Ireland would be filled with pissed off, involuntarily sterilized Protestants and Catholics while the

Middle East would be filled with equally pissed Jews and Moslems."

Silence.

"OK people, this was a very productive first meeting of the think tank," I said. "I have taken some notes, and I want to summarize the potential solutions we have discussed.

"Current estimates predict that the human population will peak between eight and ten billion sometime in the next fifty years. The longer our population remains bloated, the worse the damage will be in the form of extinction and environmental destruction. The environment has to be protected from us for an undetermined amount of time. In order to protect the biodiversity we must reduce our numbers and while we do so, we need to:

A) Facilitate development of the TIFIC.
   1) Bob: get some informants into the pharmaceutical industrial complex.
   2) Sarann: create a non-profit organization with the focused goal of promoting the TIFIC and its antidote.
B) Protect what is left of the environment.
   1) Jesús: develop your slant on Christianity that uses respect and reverence for nature as its central thesis.
   2) Dave and Teresa: combine your expertise to find ways to protect from development large portions of the ecosystem through self-propagating free market methods."

"I guess we can call it a wrap. See you at the next meeting."

CHAPTER 4

Twelfth tank meeting

"Welcome back, everybody," I said as I looked into the five windows displayed on my computer screen. I noticed that the floral print on the curtain behind Dave matched the print on the curtain behind Teresa...most interesting. This was our twelfth tank meeting. Randy, the personal assistant I'd hired to ride herd on the meetings had proven invaluable.

"Randy? You ready to roll?"

"Ready. By the way, I love what you've done with your hair Teresa."

"Why, thank you, Randy."

"OK, let's get going here," I said to the group. "What have you got to report?"

Bob started. "We have our informant inside the S&B Corporation."

"How did you do that so fast?"

"You don't want to know the details."

This did not give me a good feeling. "Fair enough. What does S&B stand for anyway?"

"The founders of the company, Peter Sedatem and Lord Bledemdri," Bob answered.

"Sedatem and Bledemdri? Real funny, what does it really stand for?"

"I was not trying to be funny."

"Oh... well, do you have any information yet?"

"Apparently, the trials are proceeding nicely, but they have postponed moving to the next level of testing using human volunteers."

"Any idea why they did that?" I asked.

"I'm looking into it," he replied.

"OK, anything else of interest?"

"I think so. A note from my mole arrived just this morning. A top-secret team of researchers has been formed to investigate the possibility of modifying the contraceptive to enhance its profitability. I, uh, have no idea what that means however."

"I think I know," I said. "This team is probably looking for a way to circumvent paying me my patent royalties."

"I'll be damned. I should have thought of that." Bob replied.

"*Expletive deleted Dave"*...a feminine synthetic voice calmly declared.

Randy had set Dave's computer up with a slight time delay the way radio talk shows do to filter obscenities over the telephone. The voice recognition software inserts a synthetic voice instead of the usual bleep whenever it detects profanities.

I continued. "This is an interesting development, any suggestions?

"We could blackmail the bastards," Bob suggested.

"What do you have in mind? I asked.

"They have hired wildlife poachers to collect poison dart frogs by the bucket."

"Really? This may provide us with an opportunity to squelch this scheme before it gets out of hand," I said. "We just happen to have a guest today, Benjamin Pinwright, director of the S&B company. I'm sure Bob already knows him. He has been set up with his own teleconferencing system. However, there were a few technical problems. We can see him, but he cannot see us. Please remember to hold down the security button on your mike if you want to make comments to the other tank members without our guest hearing them... *Dave*."

With that said, I pulled Ben's image up on our monitors. "Welcome to our think tank, Ben."

"Yes, glad to be here?" he said, looking slightly disoriented.

"These virtual meetings take a little getting used to," I said trying to make him more comfortable. He had mistaken his microphone, which was located off to one side, to be the camera, which was actually mounted directly in front of him. Ben also did not seem to realize that the same four lawyers I'd met at his lab could be seen in a reflection in the large window behind him. Obviously there to give Ben clandestine legal advice, they were also unaware that we could see them in the window reflection.

I held down my security button and addressed the group. "Those four guys in suits you see in the window reflection are Pinwright's legal advisors. They are unaware that we can see them. Be careful what you say."

"Can you tell us how things are going?" I asked Ben to get the ball rolling.

"Well, things are going as one would expect with a project of this complexity."

"I see. I hear that you have postponed the next level of testing. Can you give us a little insight into that decision?"

"The contraceptive has some unacceptable side affects. We are attempting to remedy the problem."

"What are some of these side effects?" I asked just to watch him squirm.

My question had sent the lawyers into a frenzy of head wagging and hand waving.

"Oh, well… uh, I can't really go into them all," he replied.

"I see."

In a clumsy attempt to throw me off the subject he then asked, "So, Sarann, what have you been doing with all that money you extorted from our investors?"

"Rest assured, it is being put to good use," I replied. "Now I have a question for you."

"Certainly."

"How are you obtaining dart frogs for your research?"

"We have our sources," he said with a raised eyebrow and a guilty smile.

One of the lawyers had stood up now and was repeatedly dragging his index finger across his throat. Another had his face in his hands.

I held down my security button. "Bob, are you sure you could put enough evidence together to convict these guys for illegal trafficking in wildlife?"

"In a heartbeat."

I continued. "That's interesting to me, Ben. You should have found by now that these frogs lose all of their toxicity in captivity."

Looking physically uncomfortable, Ben stole a glance toward the pack of lawyers. "At any rate, our frog supply appears to be adequate and…"

"*Prick,*" Dave interjected.

Fortunately, I had warned Ben of Dave's Tourette Syndrome. While Ben continued to talk, I pressed the security button again. "Dave, your mike is hot and your profanity software appears to be offline. You just called Ben a prick."

"I did?"

"Not funny, cut it out."

"… at any rate, this is privileged information," Ben finished saying.

"Ben, can you imagine what would happen to the price of S&B stock if the company were to be convicted of illegal trafficking in wildlife?"

Mute as he tried to interpret the wild gesticulations coming from his lawyers, Ben finally said, "Your point?"

"Just a thought, Ben."

"Wouldn't that be rather like cutting one's own throat? He countered.

"It would not be a concern to me," I said. "I would most likely have sold all of my stock long before the shit hit the fan."

Looking decidedly worried, Ben said, "We need dart frogs for our research. We have little chance of developing a synthetic version without them."

"And I have little chance of saving the planet's remaining biodiversity without adequate funding." I replied.

The lawyers were stunned. Finally, the older lawyer walked over to face the camera. "You are proving to be a worthy adversary, Sarann. What are your terms?"

"I will provide you with an endless supply of toxic dart frogs for a reasonable fee. In return, you will sign a contract guaranteeing that I will continue to receive patent royalties for every TIFIC and antidote for a TIFIC that is sold, regardless of any modifications made to the original underlying chemistry. Is that understood?"

"It is."

With that said, I logged Ben off.

"Excuse me for a moment," I said to the group. I then called Ed Osborne, the frog breeder.

"Hello, Ed?"

"Hey, Sarann, how have you been?"

"Good. How have things been with you?"

"Can't complain really. Breeding frogs is no way to get rich though; I'll tell you that much."

"That is about to change, Ed. Can you meet me at the Starbucks on 45th at say, 2:00 today?"

"Sure."

"See you then," I hung up the phone.

"Whew! That was brilliant," Jesús complimented. "The feline has been unwrapped as they say."[32]

How anyone can speak seven languages is beyond me. I have no problem forgiving Jesús for his awkward attempts to master idioms like "the cat is out of the bag," but sometimes he would be better off just not trying.

The meeting got back to business. I hadn't heard much dialog from Dave yet. "What's new with you Dave?"

"Quite a bit, actually. Teresa and I have found 100,000 acres of forestland in Washington state that we are using as a test case for our new nonprofit organization called Ecosystem Acquisition International or EAI.

"We found investment bankers willing to underwrite the sale of tax-exempt revenue bonds that we used to purchase the land. The deal depended on a favorable ruling by the IRS. Under law, tax-exempt financing must be used for a public asset. The idea here is that the land, not the trees, would be the asset, so cutting the trees for money would be legal. We will hire a timber management company to log certain areas of the property in perpetuity for profit. Our portion of that profit will then be used to pay the debts on the bonds."

I had not the slightest idea what Dave was talking about. I bluffed, "Could you elaborate on the terms 'investment banker,' and 'tax-exempt revenue bonds,' for the members of our group who may not be familiar with these concepts?"

---

[32] http://www.vetcentric.com/magazine/magazineArticle.cfm?ARTICLEID=1860

"Uh... sure," Dave generously acquiesced. "Investment bankers are people who have access to a lot of money and know how to set up the sale of bonds. They loaned us $185 million to buy the land. In return, they will see a profit by selling our EAI tax-exempt revenue bonds to the public. Anyone will be able to call up his stockbroker and buy one of our bonds. When the bond matures in ten years, it will be worth a great deal more than its original selling price. That money will come from timber sales revenues. It is a decent investment as a tax-exempt bond. If you had to pay taxes on the gain, it would not be such a good deal, and we might have trouble selling enough bonds to pay back the investment bankers. The people who bought the bonds as an investment are taking the risk that there will be enough revenue from timber sales to pay their bond off in ten years. This is an excellent example of conservation via capitalism.[33] *Expletive deleted Dave*."

Dave's voice wasn't in sync with the movements of his mouth due to his profanity time delay software. It was like watching a dubbed Japanese film

"Thank you, Dave. I think that clarified everything nicely," I lied.

Teresa joined in. "This property abuts the Olympic National Forest. It is in no immediate danger of becoming an isolated island. There will be a one-mile wide buffer zone of uncut forest surrounding the property. Many other areas will remain uncut and will eventually become old growth forest. The remaining land will be carefully managed. A healthy forest requires an occasional forest fire. Logging, when done properly, is an alternative for naturally occurring forest fires. We will replant with native species and will leave many nurse logs and seed trees intact. Forest adjacent to streams and other bodies of water will be left untouched and there will be no clear cutting."

Dave continued. "Money talks. We made a very sweet profit brokering the deal, which we will use to set up other deals. We are already receiving inquiries from other entities wishing to preserve large tracts of land from development. If this idea

---

[33] See "Conservation via capitalism," by Craig Welch, Seattle Times dated 3/24/2002

catches on we may be able to export it to other parts of the world and preserve vast tracts of ecosystem."

It was Teresa's turn again. "Our next project involves purchasing a combination of Mississippi River flood plain and adjacent fallow farm land. We are talking to the Feds about a deal to get custody of the flood plains and have hired a real estate firm to help obtain the farmland. The flood plains should not cost anything if we can convince the government to relinquish control to us. The farms will have to be purchased outright using the profits we received from brokering the Washington state deal. We intend to use the tax exempt bond idea again. There are not enough mature trees to log for profit. In this case recreational fees charged to hunters, hikers, and fishermen will pay for the bonds.

"The outer ring of the preserve will have some excellent fishing holes and hunting areas. The inner sanctum will be a source of wildlife for the outer ring that buffers the sanctuary from humanity. Wildlife populations in the inner rings will increase without human interference and will be forced to migrate outward. They do not have anywhere to go since a sea of humanity encircles this preserve, like all others. The wildlife may as well generate revenue from hunting licenses as opposed to becoming roadkill. This will also help garner support for the project from local hunting and fishing organizations. However, it will undoubtedly cause us problems with PETA. We will have to walk a fine line. The fee charged for access to the wildlife-rich outer ring will be significant. Permits will be handed out using a lottery system similar to that used by fish and wildlife departments all across the country. We will dismantle the electrical power grid in the area, and tear up and block off most of the roads."

Bob stubbed out his cigarette. "I have several ideas you can use to secure this area from unwanted trespass. A friend of mine in Texas owns a very sizeable ranch. His property is sprinkled with concealed solar powered intrusion detectors. These are inexpensive devices consisting of an ordinary car battery, a small solar panel, and a pair of motion sensors set up in series. Each one is connected to a cell phone capable of sending digital images. Being a retired engineer with nothing better to do, my buddy designed and built these things himself."

"This may be a stupid question, but what does it mean to have sensors in series?" I asked.

"It's a way of eliminating false alarms. If one sensor is inadvertently tripped by a bird or the wind, nothing will happen unless the other sensor, located a few feet away, also gets tripped within five seconds or so, indicating that someone has actually walked past both sensors."

"I think I get the concept, please continue," I said.

"People are naturally attracted to certain areas—ponds, abandoned roads, scenic views, and old farm buildings. These areas are where most of the sensors are located. When a sensor detects an infrared signature signifying that a warm body has just walked by, it calls the local volunteer fire department on a cell phone. The person receiving the call knows which detector it is just by looking at the caller ID window on his cell phone and a map with all of the other cell phone numbers inscribed on it.

"Once it's confirmed that there are intruders, the volunteer firefighter calls the sheriff and then heads for the airport to meet the local helicopter crop duster pilot. They fly to the area of trespass and escort the violators off the property to the waiting sheriff. It's all very adventurous and exciting. There is a bonus paid for every successful intruder interception. The trespassing fines offset the costs of protecting the property."

"*Expletive deleted Dave.*"

"I thought you would like that aspect, Dave."

Jesús joined in. "It is interesting to note, Sarann, that this is an excellent example of channeling natural human desires and drives. The system is self-perpetuating per your maxim that any solution must be compatible with human propensities. Chasing trespassers is fun."

"That's right," Bob exclaimed, "They practically fight over the chance to carry a cell phone and map. All this without salaries or overhead."

Teresa looked to her left and declared, "I like it."

"Me too," Dave said, looking to his right.

Bob cleared his throat. "I have a plan I'd like to discuss."

"Go right ahead," I demurred.

"Sometimes inaction can be more dangerous than action. I would like to draw an analogy. In 1941, most Americans didn't want to get involved with another European war. It is suspected in some circles that the attack on Pearl Harbor was allowed to happen to motivate the American public to accept a war. The attack stirred our deepest instincts of group protectiveness and we enthusiastically went to war. An excellent example of using primal drives and instincts to motivate an entire country to act as a single monkey troop."

Teresa interrupted. "I love a conspiracy theory just as much as the next person, but do you have a point here?"

"My point is this. If humanity is not motivated to act now to preserve what is left of the Earth's ecosystem, there is not going to be anything left to protect. Why wouldn't it be morally acceptable to forcefully protect what is left of the Earth's ecosystem?"

"Hold on there a minute," I said, "are you proposing that we lobby the government to go to war to protect critical habitat?"

"No. I am merely proposing the establishment of a large corps of volunteer park rangers available free of charge to any government or conservation organization that requests help protecting critical habitat anywhere around the world. Save the remaining critical habitat the same way the Colombian forests have been inadvertently saved by drug cartel militias.

"The structure of such an organization could emulate that of the Peace Corps, depending on volunteers for most of its workforce. These rangers wouldn't look like your standard Peace Corps cherubs. Think of a UN peacekeeping force made up of trained forest rangers instead of heavily armed guys wearing baby blue helmets. They would have to be tough and well trained, willing and able to perform all of the tasks required to protect ecological preserves in far away places. They would be equipped with the very latest technology camping and survival gear, global positioning, communications, and video equipment. Weapons would be carried by some for protection. They would be the best of the best of the best and there would be great status in being selected. Officially, their job would be to educate, patrol, observe and report transgressions to the hosting country's park security who would be in charge of confronting and possibly arresting or disarming transgressors. They would, of course, have to be invited by the citizens of the country needing assistance.

"Like it or not, competition and conflict are a part of the human condition, we may as well accept that and get on with it. With a human population this big, the problems facing the planet's biodiversity are insurmountable. It is time to create defensive perimeters and hold on."

Teresa interjected, "There are those who say we must learn to use our forests in a sustainable manner in order to save them; use the forests or lose the forests."

"Bullshit. Declare them preserves and ring 'em with armed forest rangers," Bob countered.

"Where would you get all of these volunteers?" I asked.

"Young men clamored to participate in the Civil war and both World Wars. Whatever primal instinct that motivates young men to rally to a cause will be at work for this one."

"OK," I said, still a little off balance. "What is the name of your proposed… "

"Ecosystem Preservation International or EPI," he said before I could finish. "In support of this concept, I am on the verge of submitting a book for publication that just happens to be a swashbuckling adventure story about a rag–tag group of intellectuals struggling to save the Earth's last remaining critical habitats."

"What's the title of your book?"

"Poison Darts."

Temporarily incapacitated by a sense of déjà vu, I paused, sipping my coffee.

Finally, I asked, "Does this book have a happy ending?"

"Yes and no," he replied.

Jesús cleared his throat. "Your plans may intertwine very nicely with my own, Robert. May I call you Robert?"

"Sure."

"I am building a television network to promote my slant on Christianity, one that is much more respectful of the planet God has provided us. I might be able to provide your rangers with an ideology tailor-made for the cause."

I looked into my monitor at everybody looking back at me in silence. "I'm sure you all realize that five people and a lot of money will not save this planet's ecosystems and the human race from mutual destruction. My goal is to plant a seed here. It is my wish that each of you will eventually become an independent leader of your part of the tree that will grow from this seed.

"We are attacking the problem by channeling human desires. I am not attempting to attack it directly through political means—a strategy that has so far proven inadequate. Most of the burden for protecting the planet's biodiversity is now being shouldered by giant NGOs like Conservation International, and the Nature Conservancy. I want to facilitate successful large-scale movements that will take on a life of their own. There is nothing that would please me more than to see Jesús' vision of a religion that includes a reverence for the biodiversity of the planet, to see Amazonia, the Congo, the Siberian wilderness, Antarctica and many other parts of the world roped off from further development and respected with religious fervor. In addition, if we can help people to avoid virtually all unplanned pregnancies with the TIFIC, then the population crisis will be on its way to closure. You are on your own now. Good luck. Keep me informed."

CHAPTER 5

Apparition

A few years have passed since those first tank meetings. We are finally making some progress. I was sitting in my study with a hot cup of coffee, listening to the Seattle rain beating on the skylight. I have managed to continue practicing medicine, but that will have to end soon. My life now revolves around the members of the think tank and their many activities. A news-magazine has just arrived in my mailbox, which is unusual because I don't subscribe to one. The cover photo looks like the dust jacket for a sci-fi novel. Standing in the jungle, wearing what can only be described as a battle suit, replete with horns protruding from the elbows and knees, is a camouflaged apparition from Hades. The photographer's reflection can be seen in the silvered visor of the suit's helmet. The caption below the picture reads "Eco-warriors of Colombia."

As it turns out, these eco-warriors are the security personnel of a newly formed nonprofit organization called Ecosystem Protection International or EPI. They are under contract to secure and protect the property holdings of yet another newly formed non-profit called Ecosystem Acquisition International, or EAI. The Colombian division of EAI has purchased vast tracts of undeveloped jungle throughout Colombia.

This property has been in the hands of drug cartel paramilitary forces for decades, and they have not relinquished control of it easily.

According to the article, EPI security forces are best described as extremely well equipped, superbly trained forest rangers. They use sophisticated motion sensor devices with digital cameras integrated into their design for surveillance of trails and roads. These devices also contain explosives that can be detonated only with a coded, remotely controlled signal—a design feature that discourages the drug mercenaries from seeking them out. Understandably, they are now hesitant to patrol their old strongholds. Conventional landmines are taboo. EPI is adamant that there will be no landmine victims in this conflict, innocent or otherwise.[34]

The newsmagazine's military expert speculated that the eco-warrior's battle gear is likely powered by fuel cells. The warriors probably live in air-conditioned comfort. They should be able to walk to the North Pole without feeling a chill or hike the Sahara desert without breaking a sweat. He went on to explain that, unlike space suits, which are designed to keep astronauts alive in hard vacuum at absolute zero (-459°F) these suits were designed to keep their occupants comfortable in environments that have a range of about plus or minus 100 degrees. They also did not look anything like space suits, being much thinner and lighter.

He further speculated that they employed body armor[35] on the outside of the suits when necessary and used the latest communications, night vision, and sound enhancement technologies. These lightweight, fully self-contained suits were a stark contrast to the bulky field gear and 80-pound backpacks that burden conventional soldiers. The eco-warrior's weapons were equally unconventional looking—like a miniature gattling gun wrapped around a video camera.

Part of the EAI strategy is to generously compensate farmers to abandon their coca farms. These farms are then burned to the ground to let the jungle reclaim them. Lumber and road building operations have also come to a complete halt on EAI holdings.

It was time to make a telephone call.

"Hey, Bob, Sarann here, long time no see. How are you?"

"Very well," he replied.

"Where are you?" I added.

"I don't think you want to know."

"Fine. I just read an interesting article about eco-warriors in South America. I don't subscribe to a newsmagazine, but one just happened to appear in my mailbox today."

"You don't say."

---

[34] http://www.oneworld.org/guides/landmines/stats.html

[35] http://people.howstuffworks.com/body-armor2.htm

There was an extended pause. Bob's penchant for secrecy wears on me sometimes.

"The article says that these eco-warriors are making war on the Colombian drug cartels. Why are they going after the cartels?"

Bob coughed, "Well, as I see it, every modern conflict is a combination firefight and public relations campaign. EPI is attempting to secure a positive public image. They are hoping that by subduing the cartels, they will be looked upon in a positive manner by the industrialized world. They want to be seen as the good guys. The very fact that the article referred to us as 'eco-warriors' instead of 'eco-terrorists' was a huge victory."

"Us?" I asked.

"Pardon?" he said.

"You said the article referred to 'us' as eco-warriors."

"No, I didn't."

"Yes, you did."

"No, I'm certain I did not."

"Whatever. I think I'm getting the picture here. You might have kept me better informed about your progress to date. How might I arrange a visit with these…eco-warriors?"

CHAPTER 6

The Colombian contract

As I stepped off the plane in Bogota, I remembered how bad tropical heat and humidity could be. Sweat was running down my forehead and into my eyes as I lugged my bags across the tarmac.

"Sarann?" A voice behind me sounded.

I turned and saw two Latino men.

"It is a great pleasure to meet you," the larger of the two said as he gestured toward a rusted-out wreck of a car.

I threw my bags in the trunk, tied the rope that kept it closed, and climbed in the back seat. Much to my relief, the cool breeze of a tropical thunderstorm blew through the window as the car pulled away from the airport.

The two young men sent to retrieve me spoke very little English. My attempts to make conversation failed miserably. We drove for the next few hours in the kind of downpour that can only be seen in the tropics. I like thunderstorms. I find them both refreshing and exciting. Soon we broke out into the tropical sunshine again and the car instantly turned into an oven. The jungle on either side of the road had grown thick. We turned off the main thoroughfare and after a few bumpy miles we stopped beside a shack. It was almost nightfall. I got out, grateful to stretch my legs. Suddenly the car did a U-turn and sped away, abandoning me in the middle of nowhere. This was not good. The mosquitoes were whining around my sweat-drenched face, and darkness was descending fast.

I turned back toward the jungle, and what I saw took my breath away. There, floating in mid air was a silver orb containing my own miniaturized reflection. It moved. I realized then that I was staring into the visor of some kind of helmet. I next realized that this helmet was attached to a suit, its camouflage blending almost seamlessly into the jungle background. The alien figure wearing the suit reached up and opened the visor. It was Bob grinning ear to ear.

"Good to see you again," he chimed.

"Am I ever glad to see you!" I replied, much relieved.

He beckoned me down a trail into the damp, bug infested undergrowth.

"What about my backpack?" I asked.

"Already taken care of," he said.

I looked back to where it had been.

We stopped a short way down the trail in a small clearing. Hanging by the neck from a nearby tree was the limp body of a dead space traveler.

Bob beamed, pointing at it. "There you go. Your very own eco-warrior battle outfit, or as we prefer to call it, a Sentinel suit."

"You will also need an alias. It isn't safe to be called by your real name while you're here. Pick the name of a close relative. It will be easier to remember, and you will respond better when someone calls that name out."

"OK. I have a brother named Patrick."

"That will do."

Bob introduced me to the rest of the group as Pat. Everyone had a nickname. I met Prolixity, Oscar, and Einstein.

It took a while, but with lots of help from Bob and two others, I got the thing on and fired up. The suit and helmet were both unbelievably lightweight—not at all what I was expecting. Next came several hours of training. I had to learn how to operate the communications system and a myriad of other functions. It was a marvel, comfortable, flexible, and extremely light—like a pair of flannel pajamas with footies.

Bob briefed me. "The suits are made of three separate layers of material. The inside layer is a fabric woven from carbon fiber filaments that can be heated by passing a low voltage electrical current through them. It has the feel of felt. The next layer is a shiny reflective material similar to the Mylar ponchos one finds in survival kits. This layer is primarily for insulation. The outside layer is a waterproof rip-stop material. Rip-stop doesn't describe it adequately. Actually, it has the tensile strength of chain mail armor. We kept the suit this light and flexible by using ultra-thin materials reinforced with spider web."

"Spider web?"

"Spider web, pound for pound, is about five times stronger than steel.[36] I poured a little capital into a company that was doing research on the subject. They were getting close and my funding was the breath of life they needed. They found a way to synthetically duplicate spider webs and scale them up to be used as clothing thread. We have a deal for now. We have exclusive rights for a few years before they unleash the technology on the market. Let me know if you are interested in some stock. Our suits are the result of their R&D, which brings me to the subject of body armor. The helmet contains none. Try not to get shot in the head. The design parameters included maximizing comfort and visibility, while minimizing weight. Armor had to take a back seat.

Our body armor weighs 75 percent less than that used by the U.S. military—the result of a deal with a manufacturer of body armor and the spider web company. It is segmented for flexibility and is worn on the outside of the suit. Our torso protection can stop a 45-caliber bullet without so much as a bruise. We have armor for shins and thighs as well that will keep you alive, although a large caliber round will most likely bruise you severely. You won't be wearing any armor unless we get into a pickle of some kind during your visit."

"That's good to hear," I said while hefting someone's torso armor. You would have thought it was made from Styrofoam.

"The suits are powered with electricity generated by fuel cells. Fuel cells power space stations. They will probably power our cars someday. One major technological hold-up for the auto industry—other than cost—is that fuel cells use hydrogen, which is extremely explosive. That isn't a problem with a military force, because just about everything is explosive."

The sun had completely set by now. A flash of lightning suddenly filled the sky. The other rangers, illuminated for a split second, were standing silently in a circle waiting for my briefing

---

[36]http://abcnews.go.com/sections/scitech/DyeHard/dyehard030904.html and http://www.usethesource.com/articles/02/08/21/137259.shtml

to end. Bob and I still had our helmets off. Large drops of rain began to pelt down. The briefing was delayed long enough for Bob to show me how to install my helmet and use its communication system.

As the visor slid shut, I found myself fully enclosed in a Sentinel suit for the first time. The sound of the rain and thunder were suddenly muted. The gently glowing displays inside the helmet cast an eerie light.

Bob continued over the communication channel, "A tank of liquid hydrogen contains more energy than any other liquid fuel. Two tanks the size of thermos bottles will generate enough electricity to power one of these suits for weeks. You would have to carry a hundred pounds of lead acid car batteries on your back to get an equivalent amount of power.

"You will note that the tanks are covered in protective armor. Having a bullet pierce one of these tanks would not be a good thing, although they are not under pressure so it would be less spectacular than you might think. All the same, there is a lever that will jettison the tanks in an emergency."

Bob's briefing was temporarily interrupted by some piece of urgent business. I mulled over what little I knew about super-cooled gases.[37] Making liquid hydrogen is not easy. It has to be cooled to within a few degrees of absolute zero (-459°F). It is so cold that it will actually freeze air—not just the moisture in the air, but the oxygen and nitrogen as well. It then has to be kept in a low pressure, highly insulated tank that is open to the atmosphere. This is necessary to prevent terrific pressure from building up as it warms and converts to a gas. If unused, it will evaporate from this container at a rate of about 2 percent a day. What appears to be steam coming from the space shuttle fuel tanks is actually liquid hydrogen evaporating. The temperature of this "steam" is still about a hundred degrees below zero. In a lab, the container used to hold super-cold liquids is called a Dewar Flask. It is essentially two thin-walled bottles nested one inside the other, and sealed together at the neck. The space between the bottles is a vacuum that prevents conduction of heat. Bob obvi-

---

[37] Absolute Zero and the Conquest of Cold" by Tom Shachtman

ously has come up with a spill resistant version of the Dewar Flask to store his hydrogen.

The only thing on Earth colder than liquid hydrogen is liquid helium. It has some very strange properties. For example, it has almost no viscosity or friction. I once knocked over a container of it in a lab when I was in medical school. I watched in horror as the steaming mass flowed effortlessly across the room and up the wall until momentum and gravity caused it to flow back down, across the room again, and up the opposite wall. We had to evacuate the lab until it evaporated because it was quickly spreading across the entire floor. My professor was not pleased.

Bob returned.

"What was that about?" I asked.

"A sensor has detected a group of people in the area. I don't know who they are yet. We will have to remain vigilant. Anyway, as I was saying, heating or cooling one's hands, feet, and head is normally all that's necessary since they act as the body's main heat radiators. That's why you can sit in the baking hot sun for hours if your hands or feet are immersed in a cold swimming pool. The human brain uses about as much energy as a 100-watt light bulb. This generates a lot of excess heat. The reason scalp wounds bleed so profusely is because the skin surrounding the skull is highly vascular. Most of these blood vessels are there to carry waste heat away from the brain to the surface of the skin where the heat can be dumped. From an engineering perspective, the human skull is an armored cooling jacket.

"It takes very little power to control the temperature of the suit clearance volumes around your feet, hands, and head. These spaces combined are no bigger than the volume of a coffee mug.

"During periods of high physical exertion, the suit also begins cooling the air that's pulled into the lungs. The most efficient heat exchanger on the planet is a healthy pair of lungs. Therefore, the most effective way to cool the human body is to breathe cold air. Thirty-eight-degree air going into your lungs comes out heated to about ninety degrees."

Bob was called off yet again. His briefing had contained a little more detail than I cared for. I was standing in a hot, humid, insect-infested jungle, in total air-conditioned comfort, and that was all I needed to know. No more sweat, humidity, or bugs.

The night vision system was amazing. We were a group of four individuals, standing in a dark jungle, chatting with each other via a sophisticated communications system. Our voices could not be heard outside our helmets. To a bystander, it would look like camouflaged space men having telepathic conversations in total silence. I was especially impressed with the built in global positioning system. Our precise location on the planet and relative position to each other could be projected right onto my visor just like a jet fighter's heads-up display. Many high-end luxury cars have similar systems built into their dashboards.

I opened my visor and looked around. "Do you guys have a camp nearby?"

"We don't do camps," Oscar replied.

I was perplexed. "Don't we need tents and sleeping bags and cooking kits and all that stuff if we're going to be hiking in the jungle for a couple of days?"

"You will find that your suit is more comfortable than a sleeping bag. It has inflatable padding built in at all the right places. When it's time to sleep, go find yourself a comfy spot, inflate the padding to suit yourself, and just lie down. There's a blow-up pillow in the lower left pocket. Your other storage pockets contain enough dehydrated food to last you about two weeks. The water filtering kit will make any source of water you find perfectly safe for consumption."

"Why do they call you Oscar?" I asked.

"I can get pretty irritable."

"You're a grouch."

"Exactly."

"How about Einstein over there?"

"Mr. Know-it-all."

"I see."

How about Prolixity?"

"It was better than calling him motor-mouth."

I had a lot to learn. I would be a liability to these people for some time.

Bob returned. "We are presently on EAI land. Officially, our job is to protect it from trespass. From here, we're hiking to our main base. This should give you a feel for our operation. We will cover ten miles tonight before we rest. Sar…ah, Pat will be in position number 3. Prolixity takes point. Ten meter spacing, laser communication only. Enjoy your walk, everybody."

I experimented with the sound mod system as we hiked, filtering the jungle noises that were outside my suit. I was able to attenuate the noises made by the bugs and frogs, while at the same time accentuating the sounds of our own voices and footsteps.

I found that walking through a deathly silent jungle was just too creepy. I turned the system off. A few minutes later we frightened a squawking parrot off its nighttime roost.

"*Remarkable bird, the Norwegian Blue, isn't it, aye? Beautiful plumage!*"

"Norwegian Blue?" I blurted over the radio network.

"Did you hear something, Pat?" Oscar asked, sounding concerned.

I realized then that the voice had been too clear, too close. It hadn't been real. My slight propensity to hear voices—particularly Monty Python voices—that aren't really there had caught me with my guard down.

"Sorry, Oscar. Just my imagination," I half lied.

The excitement and its accompanying adrenaline had me wanting to jog. Fatigue finally set in as the night wore on. At one point we had to ford a large stream. The suits were waterproof and even had an emergency oxygen supply, so drowning was not a big concern. They tied a rope to me just in case.

A couple of hours before sunrise another heavy rain began to fall. I spotted a glowing shape with my night vision system, crouched low, about forty yards to our right. I shot a message to Oscar using the line-of-sight communication laser from my helmet to his.

"Oscar, do you see that shape off to our right?"

"It's a jaguar," he casually replied.

The hair on my neck stood straight up, the same response a rat has when confronted by a cat. I knew a few things about jaguars. Most great cats kill with a bite to the throat. Jaguars kill their prey by biting right through the skull.[38] A big part of their diet consists of river turtles. They eat them shells and all. Although similar in appearance, jaguars are much more powerfully

---

[38] http://users.netropolis.net/nahury1/jaguar.htm

built than their African and Asian cousins, the leopards. Oscar hadn't sounded very concerned. I was probably the last member of the group to see it.

My thoughts were interrupted by a warning beep from my sound amplification system. The voice recognition software had detected unfamiliar voices nearby.

"Everybody freeze," Bob's voice blared through my headset. "Somebody deploy a parabolic microphone."

I waited in silence.

"They're off our 120 radial," another voice piped in.

"Sit tight, Sara… er, Pat. We'll be back soon."

Bob was having a harder time with my alias than I was. I hunkered down in the pouring rain to wait. My radio was deathly quiet. The group must have gone to a secure frequency.

Less than an hour later they returned to fetch me and we continued our journey into the night. Finally, just before sunrise, we stopped to sleep through the day. I had to satisfy my curiosity first so I cornered Bob again.

"What happened at that camp last night?"

Bob was looking as tired as I felt, but he accommodated me with an answer. "They had dogs with them which means they were poachers, probably after the jaguar we spotted. We disarmed them, relieved them of their tents, shirts, shoes, and dogs. Without the security of their weapons and dogs, and with the insects trying to eat them alive, they will not be sleeping well until they get back home."

With that said, he assigned the first shift of sentry duty to Prolixity and moved off to take care of perimeter security.

I could see that a ranger in a Sentinel suit would have no problem running down a fleeing poacher. The suits were designed to maximize freedom of movement. I found myself a relatively level patch of ground and settled down to get some rest. Bob was right. The suit made a very comfortable sleeping bag. The inflatable padding allowed me to adjust it to fit my body's needs. I also learned that I could adjust the light being filtered by my visor. I set the suit's temperature to a comfortable 68 degrees and slept like a baby.

I awoke to a scratching sound. I looked to my left and found myself staring at the undercarriage of a giant insect attempting to scurry over my visor. I must have screamed because two rangers were beside me instantly.

"Sorry about that, guys," I said, apologizing for my embarrassing vocalization.

A Hercules beetle had been attempting to traverse my head.[39] I was familiar with this species of insect. Bigger than my outstretched hand, they are fairly common in this part of the world. Although scary looking, they're actually quite harmless. I picked it up for closer inspection. A friend of mine once made the mistake of letting one crawl up his bare arm. These beetles have a powerful six-legged grip that can be very uncomfortable. There is no way to pull it off once the climb up your arm has started. You just have to grin and bear it until the beetle gets to your shoulder. You have to make your move at that point before it gets into your hair.

It was late afternoon already, and I was famished.

"What's to eat?" I asked the crowd that had gathered. Apparently Oscar had chaperone duty for the day and came over to educate me.

"First thing you need is water. We brought some into camp this morning. It's over there in that jug."

I sauntered over to where he was pointing and looked into a plastic jug full of muddy water and wriggling mosquito larvae. "You're kidding, right?"

"You have to filter it first. Here, I'll show you."

He pulled a device out of one of my pockets and plugged it into my suit's power supply. In a few minutes, I had a quart of crystal-clear hot water. He then pulled a couple of food packets out of another pocket and showed me how to turn them into something edible. I had a soy version of corned beef hash followed by a helping of oatmeal and brown sugar. It was delicious.

"Why don't you guys use iodine to sterilize your water? Wouldn't that be less complicated than a filtration system?"

"Not really. There are bugs in the water here that iodine won't touch. Besides, the water has to be hot to work with most of our dehydrated food and campfires are verboten. If we are going to pass the water through an electric heater, we may as

---

[39] http://www.ivyhall.district96.k12.il.us/4th/kkhp/1insects/hercbeetle.html

well filter it while we're at it. We have power to spare with all of this liquid hydrogen."

I tried to stay out of the way as the group went through their camp breaking rituals. Every packet was put back into the suit pocket it came out of. All signs of our stay were removed. The various motion sensors and listening devices that had protected our perimeter through the day were retrieved, and we were off again. I had a lot of questions to pester these guys with on tonight's hike, and since Bob had delegated Oscar to baby-sit, it was his turn to be tormented.

"Oscar, what did you guys do with the poacher's dogs last night?" These Sentinel suits were amazing. I was talking to a guy 30 feet behind me like we were face to face.

"Well," he replied, "we used a small caliber weapon that has a silencer. It's a fairly humane way to put a dog down. They never knew what hit them."

I was stunned. It hadn't dawned on me that they had killed the dogs.

"Our primary small arms weapon consists of five small caliber gun barrels wrapped around a camera lens. Like everything else we use, it was designed under Bob's direction. With today's optical systems, the idea of having to aim your weapon by holding it against your shoulder while exposing your head to enemy fire is rather obsolete and stupid to boot. With this weapon you stay hidden behind a nice solid tree or brick wall and hold it out to look for targets. It has a powerful digital telescope, a laser range finder, and infrared night vision capabilities. You can deploy its tripod and operate it remotely from 150 feet away if you like.

"The ordnance is basically a rocket-propelled, armor piercing bullet. The technology was first introduced in the Vietnam War. They didn't work very well. They accelerated too slowly for close-in fighting and were inaccurate at long range. Bob knows a munitions expert who fixed both problems but couldn't interest the U.S. military because they have too many weapons systems to deal with already. The bullets now have two stages. The first stage sends them out of the gun spiraling like a bullet. The second stage fires for only a millisecond part way through its trajectory, thus greatly enhancing its range. If they hit something before the second stage goes off, it acts like an ordinary bullet. An impact disables the second stage rocket. Unguided rockets are usually inaccurate because the thrust vector causes them to yaw and pitch. This second rocket fires so accurately and for such a short period that it does not alter the trajectory.

"There are three main advantages to the rocket propelled bullets. One, they are extremely accurate at long range. Two, they don't have a recoil. Thus, we can fire them remotely from a tripod without having 'em tip over. Sort of like a miniature version of the recoilless cannon used by the Marine corps. The third advantage is that because the gun barrels are actually thin-walled rocket tubes, they don't weigh much."

"What do you need armor piercing bullets for?" I asked, ducking under a liana vine.

"These weapons are not designed for shooting people. They are for shooting equipment, trucks, cars, bulldozers, helicopters on the ground, anything with an engine. The bullets go through engine blocks like they're made out of butter."

His reply relieved me of a great deal of anxiety.

"It can be programmed to fire in several different modes, automatic, semi-automatic, starburst pattern, and ambush mode. When you fire it in the star pattern, it shoots five rounds in about one quarter of a second. This means that, essentially, they are fired simultaneously. The fifth bullet arrives on target one-quarter of a second after the first bullet. If you're a truck on the receiving end, it feels like they all arrive at the same time. This feature makes it a lot harder to miss, kind of like a shotgun.

"It can also be tied into the battle computer for ambush mode. Here's how it works. Say there are ten of us aiming at a cartel convoy. We each acquire a target, a tire or engine compartment, and hold our trigger down. The gun will not fire however until at least seven—or whatever number the commander picks—of the ten triggers are being held down. When that last trigger is pulled, all seven weapons fire simultaneously, and the trucks stop moving. If you combine the star burst mode with the ambush mode you can just imagine what it would be like. A wall of steel screaming 600 miles an hour toward the critical parts of a vehicle."

"How did you sneak up on them with all of those dogs around?"

"Not a problem really. It was raining hard, perfect conditions for a close-in raid. We can move about the jungle without being detected because of all the background noise. Our ability to filter out background noise while amplifying other sounds gives us a big advantage. We could hear every word being spoken in the camp as we approached."

I thought about all this for a while before I came up with my next question. "What would you have done if there had been a sentry posted?"

"There was a sentry. We located him with our infrared sensors."

"What did you do with him?"

Bob's voice came through the head set, "I'll field that one, Oscar. You're asking some questions you may not like the answers to. Fortunately, he was asleep. The outcome would have been different had he been awake and started shooting at us. Colombian law enforcement authorities have officially deputized every member of the EPI field staff. We are authorized to use deadly force in self-defense."

"How did you manage to get everyone deputized?" I asked, strongly suspecting that the answer would be bribery.

"I think you know the answer," Bob said, confirming my suspicion with his annoying ability to guess what I'm thinking.

I had enough to chew on for a while. We continued our trek into the rainy tropical night.

In the wee hours, my COM set crackled to life. I eavesdropped.

"Clearance code please," a polite voice asked.

"Alpha-Teresa-four-bravo," Bob replied.

"Welcome home. Cleared for access to zone 3."

"Roger. Confirm party of five."

"Affirmative."

"Turn to vector three-two-zero for fifteen."

"Re-supply status please."

"Five-three-three-three-two-seven."

"Prepare to receive encrypted burst... three-two-one-zero."

We walked on in silence for another hour or so, and then I practically ran into Bob who was supposed to be ten yards in front of me.

"What's up?" I asked.

"We have arrived."

It didn't look like we had arrived anywhere; just more jungle with trails going here and there. After two nights of hiking through jungle, it was all a little anticlimactic. I settled in to rest and wait. The sun was just rising. I fiddled around with my suit controls and found the television. Just as I suspected, over 200 satellite channels to choose from and nothing worth watching.

I must have sat on an ant nest because I could see them running across my visor. They seemed larger than normal.

Several other rangers soon joined me. Their visors were up and a nearly transparent bug netting resembling camouflage panty hose had been pulled over their helmets.

"Getting some rest?" Bob asked in a wry tone of voice.

"I am, thank you very much," I retorted through the ever-growing swarm of ants on my visor.

"You do realize that you're sitting on a bullet ant nest," Bob said, unable to contain his grin.

"Bullet ants!" I yelled as I reflexively leaped to a standing position. No wonder they seemed bigger than usual. Bullet ants are huge, over an inch in length.[40] They have monstrous mandibles and a sting that has been described as the most painful known to man, twenty to forty times more painful than a wasp sting—like being shot by a bullet. Three to five stings are enough to kill a person. Although my suit protected me, I decided to sit someplace else. Better safe than sorry.

Bob continued. "I'd like you to meet some of my eco-warriors, or as I prefer to call them 'field managers'".

As Bob was talking, I noticed one last bullet ant walking across my visor. I reached up to knock it off and realized to my horror that it was on the inside.

Bob ducked as my helmet flew past him. What did I tell you about keeping your visor locked while you're resting?"

---

[40] http://www.sasionline.org/antsfiles/pages/bullet/bulletbio.html and http://us.expasy.org/spotlight/articles/sptlt014.html

"Won't happen again," I said.

Bob and I were the only Americans in the group. Everyone else was Colombian. Introductions were done. Some pleasantries were exchanged and then one by one the young men excused themselves, leaving just Bob and me in the small clearing.

Bob pulled out a bottle of Jack Daniel's and offered me first swig. I declined.

"What socioeconomic class spawned you?" Bob asked out of the blue.

"Raised on welfare by a single mom." I replied.

"Really…" Bob mused.

"Yeah, we lived in a bad part of town. I'll bet that a third of our peers were either dead, in prison, or pregnant by the time we got out of there."

"How did you manage to pull yourself out?" Bob asked.

"I'm not sure I have," I concluded. "I'm no longer impoverished but the environment I was raised in is still a part of me."

"In what ways?"

"I know the essence of poverty. I've come to the conclusion that luck has more to do with our everyday lives than any other single factor. I was lucky to be born into a society that gave me the opportunity for an education and a chance to improve my lot in life. Had I been born to a street person in Bangladesh my life would look very different."

"Yes, we tend to give ourselves too much credit… human nature," Bob concluded.

"Human nature," I concurred.

"How close are we to your main camp?" I asked, changing the subject once again.

"This is it. Our group is dissipated over a large area. Because the suits are fully self contained, we don't need a mess hall to prepare meals or tents to sleep in. I know the location of every ranger via his global positioning transponder. I can talk one-on-one with anyone in this contingent at the touch of a button. They are in their assigned positions and can be sent to a new position in a heartbeat."

"Where did you come up with the idea for the Sentinel suit?"

"Well, the idea came to me one cold winter day when I was flying my ultralight. I was freezing my ass off and wondering if anybody sold electrically heated clothing for idiots like me. I looked around and found several manufacturers of heated clothing for motorcycle riders. The technology is the same as that used for electric blankets. It uses a lot of power and has to be plugged into the motorcycle's electrical system—primitive stuff.

"Then I stumbled upon a jacket that uses a radically new kind of heated fabric technology. The fabric is made of electrically conductive carbon fiber filaments.[41] It requires very little power and runs on batteries sewed into the lapel. The process used to make the fabric is still a secret. I didn't need to know how they make it. I just needed to be licensed to use it.

"I made a prototype for my personal use. It worked well. When I measured the amount of electrical power it was using, I realized that a small fuel cell could power the thing in subzero weather for weeks on end.

"My ultralight was already using a fuel cell for electrical power. The next step was to integrate one into the suit's design. I then removed the global positioning system from the ultralight and integrated it into the suit. I owned a cell phone at the time that was voice activated and capable of sending and receiving video images. I thought, 'what the hell,' and integrated that into the suit as well. The cell phone was quickly followed by the addition of a 700 X digital zoom lens with night vision capability, both pirated from my $500 video recorder. Ultralights are notoriously loud so the last modification to my helmet was a dynamic noise cancellation system taken from ear defenders used at gun ranges.

"That was pretty much it until summer rolled around. Then I had to find a way to keep cool. I found the answer to that problem in the form of a battery powered beer cooler at my local sporting goods store. These beer coolers incorporate solid state heat pumps called TECs (Thermo-Electric-Coolers) that use the Peltier effect. DC current flows through the TEC making it cold on one side. A single-stage TEC can achieve temperature differentials exceeding ninety degrees.[42] I envisioned what it would feel like if my feet, hands, and head were inside their own little

---

[41] http://www.gorix.com

[42] http://www.hitechtec.com/Applications.htm

coolers. In addition, the hydrogen gas evaporating from the tanks has to be warmed before it can pass through the fuel cell. I supplement the TECs using heat exchangers to warm the hydrogen with body heat. Essentially, I am tapping into the tremendous amount of potential energy that was created to convert the hydrogen into a liquid.

"The key here is that these suits require a constant electrical current to keep cool or warm. Technology begets technology. The technologies that allow us to make and store liquid hydrogen combined with the technology called a fuel cell makes the concept of a Sentinel suit feasible."

"Why do they have these horns on the knees and elbows?"

"Well, originally, they were an idea to increase traction when crawling up or down steep inclines. I found that they weren't really effective for that, but because they gave the suit such a formidable appearance I decided to keep 'em."

"Just incredible," I said as I sat down and removed my helmet. "This operation is bigger than I imagined. I thought you were supposed to be a bunch of forest rangers patrolling a preserve. This has the look and feel of a guerilla war."

"You're telling me. I picked Colombia for our first test. If we can successfully protect this preserve, we can protect any preserve. I don't expect every security contract to have this level of difficulty."

Bob took another pull on the whiskey bottle. "Keep in mind that we have not joined America's War on Drugs. We are merely consolidating our grip on parcels of rainforest that are now owned by EAI. We are making some income from selling confiscated drugs to the DEA. It is more cost effective for them to buy the drugs from us than to try to intercept them as they enter the country. The truth of the matter is that we have not made a dent in cocaine production here. I seriously doubt this will affect the drug supply chain. The demand will be met by other sources."

"So, where are you getting the money?" I finally asked. "Your operating budget can't be meeting this demand."

Bob coughed perfunctorily, and continued. "We have a wealthy patron."

"A wealthy patron? He must be *really* wealthy."

"He is."

"May I ask who?"

"You don't want to know."

"Damn it. Will you quit saying that? Can you at least give me a hint?"

"He doesn't spend much time in Seattle."

This was getting ridiculous. It shut me up for a minute though. At least I knew who he wasn't talking about. Well, whoever it was, I was glad to hear of it.

"OK, tell me this then. Have you actually met this patron face to face?"

"Oh yes. He spent a week in a Sentinel suit."

"Amazing, just amazing," I said. "I have to admit, that's the best recruitment card you have."

"Yes, we had to be especially careful to keep him out of harm's way while he was here," Bob added.

"What do you have up your sleeve after you tap this guy out?" I asked.

"I have something in the works that may be worth fifty million dollars if everything goes as planned," he said.

"May I ask what that might be?"

"Well, it has to do with U.S. sponsored bounties on the heads of certain individuals and the organ donor box you are sitting on."

Standing up, I said, "That brings up my next question. You must be taking casualties at this point. How are you dealing with them?"

"We have some excellent medical facilities in the field. We stabilize and then transport as fast as we can to the nearest hospital where our people and money can take care of things."

"Can I assume that the government is pleased to have you here?" I continued.

"Unofficially, certain elements don't want to see the drug money cut off. That cash lines the pockets of a lot of politicians all through South America. Unfortunately for them, they can't complain too loudly without showing their pro-drug sentiments. Unofficially, they want to get rid of us as badly as the drug militias do. It's a big secret at this point because they can't very well tell the world that they are actively trying to displace the forces that are trying to displace the drug cartels. We have avoided any serious conflicts to date.

"The tremendous range of our rocket munitions allows our sniper crews to riddle with bullet holes any aircraft we catch on our property. This includes government aircraft trying to covertly support the drug militias. We don't shoot them out of the air because that would kill somebody. We wait until they're safely on the ground. The weapons make very little noise. When the flight crews come back to do their preflight inspections, they find all these small caliber bullet holes in critical areas of the transmissions and turbines. The damaged planes and choppers are grounded for weeks or months for expensive repairs. No pilot will dare fly an aircraft with a bullet hole in the engine section because there is no telling what oil or fuel line may have been nicked. A damaged gear or turbine blade could prove fatal and the pilots know it.

"They have seen our firepower and have gotten the message loud and clear. I think it's had an impact on their enthusiasm. Our primary objective is to clear our land holdings of the drug cartel militias and their farms. If we are successful in this endeavor we are hoping to secure a positive image from the world's press organizations."

"It all looks very convoluted to me," I commented.

The whiskey was looking appealing so I accepted a swig. After the burning sensation subsided I asked, "What is your overall strategy for going after the cartel militias?"

"Well, they don't really qualify as militias. They more closely resemble a nasty bunch of well-armed thugs. We will first destroy the coca crops and compensate the unarmed farmers who are caught up in the middle of all this. When put in perspective, it isn't all that expensive. For the cost of one cruise missile, we can give fifty peasant families forty thousand dollars each. Eventually this will eliminate the militias funding. In the mean time, however, it will really piss them off. Like hitting a hornet's nest with a stick."

"Do these families need protection from the cartels after you compensate them?"

"What for? They can take the compensation or leave it. The cartels understand that much."

"I don't know, Bob, indigenous people all over the world are being displaced by rich people. This just looks like more of the same. Farmland in Nepal is around $10,000 an acre. In Tanzania, land is $25,000 an acre. Few people want to give up their livelihoods just so they can move to a big city and join the unskilled, unemployed labor pool."

Bob was silent as he seemed to mull my accusation over. "Well, if they are smart, they will use that money to start another small business, which is all a farm is anyway. Competition is the way of nature, it is also the way of human nature, and by extension it is the way of free market economies. How did these farmers get their hands on this land in the first place? They probably took it from someone else. Correct me if I am wrong, Sarann, but don't you live in an area of Seattle where the average house costs about half a million dollars?"

"Yes," I admitted.

"Then tell me, what happened to all of the poor people who used to live in that neighborhood? Have they been displaced to the seedy part of town?"

"Probably," I had to admit again.

"What do you want? Do you want the world to be one big fuzzy commune where everybody shares everything equally? Sound familiar? Have you read any Marx lately? It's been tried, repeatedly; it doesn't work. It goes against the very grain of human nature. Force a complete redistribution of wealth and the free market engine will stall, starving half the people on the planet to death. The person with the answer to that problem has not stepped forward yet."

"All right, enough," I said, quelling my own anger.

"Fuck it, Bob continued. "Everybody is involved in drug trafficking down here, from the official government to the various left and right wing paramilitary armies.[43] When our surveillance cameras catch groups of armed men cutting through our land holdings, they could be anybody, the leftist paramilitary organization FARC, the right wing counterpart called AUC;

---

[43]http://spectrum.troyst.edu/~sltaylor/interam/Colombian%20Drug%20Cartels%20Exploit%20Tech%20Advantage.htm
and http://www.rand.org/hot/op-eds/061701SDUT2.html

even the IRA has a presence. Everyone traffics in drugs to some extent, including the Colombian military and the DEA. They sometimes even collaborate with each other. We are painting a simple black and white picture for the press to disseminate for us. The good guys (us) against the bad guys (the drug dealers).

"It isn't nearly that simple of course. The drug traffickers control cocaine production from the raw material to the final product. This insures their profits when it is sold to Mexican delivery specialists. The Revolutionary Armed Forces of Colombia (FARC) makes a fortune by protecting the coca fields, guarding labs and airstrips, and even taxing cocaine production. Paramilitary groups allied with the army also profit from the cocaine trade by protecting supply routes and running labs in areas under their influence. There are more than 300 active drug smuggling organizations in Colombia today. The AUC or paramilitary membership includes cashiered Colombian army officers as well as former guerrillas. They are known for their brutality and have become involved in the drug trade as well."

"It all seems rather convoluted," I suggested.

"That is an understatement, Sarann. The easy availability of high-powered assault rifles has turned every bored, adventure-seeking young man with nothing better to do into a freedom fighter for one cause or another. This scenario has its counterparts all over the world. Beirut was a recent example of this phenomenon. Other areas of the world with chronic paramilitary problems include Africa, the Philippines, Ireland, and Bosnia. The list goes on and on. Young men are drawn by the excitement, the adventure, and the sense of belonging. It is just human nature, especially male nature. Street gangs in New York are micro versions of the same phenomena without jungles to hide in. The young men filling the ranks of the EPI security teams are motivated in exactly the same way. Their mantra is to protect mother Earth."

I interrupted. "Then what makes your group any different from these paramilitary groups?"

"It isn't," Bob answered bluntly. "The other groups are laboring under the mistaken idea that they are right and everybody else is wrong. The reality is that the group with the biggest stick is going to win. We have a very, very big stick in the form of high tech wizardry and excellent funding. Our cellular communications network, combined with a satellite-based global positioning system allows us to watch every road and trail in our area of influence, day and night. We have distributed brochures which state that anyone carrying a weapon on our property is a legitimate target. A gun in hand is like having a sign on your back that says, 'just shoot me.' We can't disarm all of Colombia, but we can and will disarm anybody who makes the mistake of crossing our boundary. Those who heed our brochure's warning and drop their weapons will leave unharmed. Those who stay to fight or attempt to escape with their weapons are another story. We are a guerilla army among guerilla armies. The various government and paramilitary groups can continue to play their war games, but they are going to learn to avoid at all costs this area of the planet we have roped off as a sanctuary."

"People are getting killed?" I asked.

"They are," he replied uncomfortably. "Those who attack us with deadly weapons are going to have to give those weapons up one way or another."

I had touched a nerve with that last inquiry. Changing the subject I asked, "Can I see how one of these digital sensor things works?"

"Surely," he replied as he pulled a camouflaged, militarily hardened laptop from a bag and plugged it into his suit's power supply.

"Here's some footage from a group that penetrated our periphery a few days ago. The motion sensors indicated activity in this area. We turned on the surveillance camera to see what was up and found these guys."

We watched a video of some hunters with shotguns moving along a trail.

"They were not worth sending a patrol to intercept, so we waited until they got near a remote-controlled loud speaker and told them to drop their guns and get off the property. We then lit off a couple of mines behind them to add to the fear factor. Detonating mines in front of or right behind intruders has never failed to send them packing. Just look at how big around their eyes are in this shot."

I realized then that Bob was fighting the first remote-controlled armed conflict. The technology was all there, computers, motion sensors, cellular communications, digital cameras, fuel cells. Bob was just the first guy to put it all together. What sane person is going to stay and fight an enemy who announces that you are in his digital camera sights and that you are standing on his remote-controlled mines?

"You must be talking about a lot of mines and sensors here."

"No, not really. They're only along roads and trails. We put them at every entrance and at strategic locations along their length. It only takes a sensor every mile or so to cover a road or trail. We don't need to know exactly where the intruders are. All we need to know is which sensors they are between on a given trail. We have very few mines. Detonating them in front of or behind an intruder takes patience because we have to wait until they get near enough to one. We still have a lot of work to do to finish securing this property. Our field personnel spend most of their time deploying and maintaining sensors and cameras. It is very rare that we have to physically engage armed trespassers."

"This is good whiskey!" I proclaimed.

"When do you engagement next anticipate?" I misspoke.

With that, Bob stood up to take his leave. Grinning, he said, "Soon enough…and may I suggest that you get some more rest."

I wasn't feeling particularly tired and decided to badger Oscar with some more questions.

"Hey, Oscar, has one of these Sentinel suits ever fallen into enemy hands?"

"It has happened. We were able to free the ranger, but the suit had to be destroyed."

"How did you manage to destroy the suit?" I asked.

He gave me a pained look. "These helmets are rigged with high explosives. They can be detonated by the commander via remote control or manually by the suit's owner. If you're ever captured it's a good idea to put a lot of distance between yourself and your helmet. An enemy force in possession of a helmet will be monitored by its radio and digital camera. We can watch and listen to everything being said within earshot of the suit. We can decide when to detonate. Optimally it will happen in a room full of curious cartel big-wigs."

This information shattered my alcohol-induced sense of contentment. "So, if a ranger gets shot in the head, he blows up?"

"Relax, this explosive requires a special high temperature ignition source. Bob once held a match to a chunk of it to prove the point."

"Thanks for the info Oscar, I'd better get some more rest."

"My pleasure."

As I lay there, I got to thinking about armed conflict. It certainly is not a modern day concept. The fossil record clearly indicates that warfare has always been a big part of the human condition.[44] Some scholars estimate that about 25 percent of all males in prehistory died at the hands of other males. There is ample evidence from fossil skull samples that right-handed men routinely hit each other in the head with blunt objects. We know this because a significant percentage of the male skull samples dating back to pre-history have healed fractures on the left side of their upper forehead. As technology progressed, clubs were replaced with blades and men began to hew and hack at each other. The fossil record begins to show cut marks on bones at this point. Next came projectile weapons, spears, and arrows. The iceman mummy recently uncovered in the Alps lived over five thousand years ago at a time when the Stone Age was just giving way to the Bronze Age. Found in his possession was a beautifully crafted and polished bronze ax, a bow, a knife, and a quiver full of arrows. At first, researchers speculated that he might have been a peaceful sheepherder who had been caught in a blizzard. The bronze ax was too soft to be useful for cutting down trees. They speculated that it must have had ritual significance. The researchers were wrong on all counts. Later exams showed that the iceman had been shot in the back with an arrow. In addition, he had defensive wounds indicating that he had been in a fight for his life. The most interesting finding was blood from four other individuals discovered on his knife. The beautiful ax was actually a deadly war club. The noble savage is a myth.

Modern war typically involves ripping your opponents asunder using high-energy chemical reactions in close proximity

---

[44] Constant Battles by Steven A. LeBlanc with Katherine E Register

(explosions), shattering your opponent's bones and flesh with impacts from high velocity projectiles (bullets and shrapnel), or incinerating them through rapid oxidation (napalm munitions and flame throwing devices).

Germ warfare is nothing new. I know of more than one instance when an ancient besieging army used its catapults to throw plague-riddled bodies over a city's protective walls. Early settlers were known to pass out smallpox-infected blankets to Indians.[45]

The Gulf War was fought for two reasons. It protected what we perceive to be *our* oil supplies, and it prevented Israel from using its *secret* nuclear arsenal. The reasons for the Iraq war still remain fuzzy. Bob's goal is to protect the remaining critical ecosystems of the world for as long as possible, a kind of desperate holding action. It is extremely rare to find a conflict being fought for such a high moral imperative. Just the same, I can't help but feel empathetic toward the drug cartel's hired guns. They are just trying to improve their lot in life. They don't use cocaine themselves. U.S. citizens want their product so badly they're willing to pay a small fortune for it. I drifted off to a fitful sleep.

I was awakened by Oscar's voice in my helmet. "Time to go."

It was dark again, and I was a little disoriented. "Which way?"

"Follow me."

I fell in line, and after walking for about twenty minutes we joined up with Bob and his command group.

"Glad you could join us," Bob said. "We're going to show you how we secure EAI land holdings. It's a lower risk endeavor than subduing drug cartel mercenaries. We call it Operation Roadkill. After a lot of research and thought, we have concluded that the best way to defend an area from trespass is to remove all roads and trails into that area. It's simple, economical and relatively non-violent."[46]

---

[45] www.nativeweb.org/pages/legal/amherst/lord_jeff.html

[46] http://nationalzoo.si.edu/publications/zoogoer/1999/5/wildlifelanes.cfm

We hiked for three hours, finally stopping alongside a large river. An EPI scout was already positioned there awaiting our arrival. Looking out across the water from our hiding place we could see a road crew about a quarter mile away putting the final touches on a major road repair. The rangers deployed along the riverbank and began setting up their weapons on firing tripods.

Bob briefed me. "When the board of directors of EAI of Colombia, staffed entirely by conservation minded Colombian citizens, purchased this land, we gained the moral high ground. We aren't foreign invaders with a guerrilla army—we are forest rangers protecting our nature preserves. We also use this property as a base of operations from which we infiltrate surrounding areas. No one is quite sure who is blowing up all the roads, but I'll tell you this; it is amazing what a few sticks of properly placed dynamite can do. The cartels spend their own money repairing roads because they can't wait for the government to do it. Without roads, they cannot transport their goods, personnel, or equipment. Watch this."

He tapped the scout's helmet who then pushed a button on a remote control device. A mountain of dirt flew into the air. When the dust cleared, two large craters could be seen where a fresh new road had been only seconds earlier. They were already filling with water from the river. The road crew and their security detail had been a safe distance away, but were now scurrying about the craters like ants.

I also noticed dozens of faint smoke rings half way to the road. "What are all those little smoke rings?"

"They were created when the second stage of the rocket bullets ignited." Bob answered.

"Don't they tend to give your position away?"

"Sure do. That is why we won't be staying long. As I was saying, we typically place charges in an area where it will be expensive to repair, near a water source or low water table that can drain into the hole. We blast a string of great big ponds along a road. It's a lot faster and cheaper to blast a hole in a road than it is to fill it back in and pave it. We have the economy of dynamite on our side. The ponds are also a nice enhancement to the ecosystem. Thousands of miles of roads are virtually impossible to protect. We take pride in our ability to blow new holes in the road just hours or days after a road crew has spent weeks repairing it. We sometimes bury a second charge in the ditches and wait until the road crews have spent a week fixing them before we blow it up again, as you just witnessed. The firing platforms were all tied into the remote control that ignited the explosive. If you look closely, you will notice that the road grader, earthmover, and supply truck are listing in our direction from flat tires. Their engines are also full of holes. The road crews are still unaware of this fact because the sound of the blast covered the sound of the bullets slapping into their equipment. Essentially, we are a band of taggers using dynamite instead of spray paint to annoy the authorities. The solution is elegant; wreck the roads into the forest preserves and keep them unusable."

"What about people who live at the other end of a road?" I asked.

"You will see how we handle that at our next stop. Come on, we need to get out of here."

We hiked another two hours or so before we came upon a coca farm. From the edge of the jungle we could make out hired farm hands going about their business and two armed guards.

"This is our property," Bob told me. "The armed guards are problematic. I suggest that you wait here for a while."

I accepted his suggestion. A half-hour later a ranger showed up to escort me to the farm. As we walked across the field, I saw the farm help collected into a group talking to Bob. There was no sign of the guards.

One of the farm workers could speak English. Bob had his visor up. "Here's a little unemployment compensation," he was saying as he handed out wads of cash. "If we see you back here, you will meet the same fate as the guards. Understood?"

The farmer enthusiastically nodded in the affirmative.

"OK, then, get yourself on the truck and get out of here." Bob finished. He then walked over to where I was standing.

"Typically, we assist them with relocation. We pay the workers about two years worth of their salary to keep them happy—our version of wealth redistribution and poverty reduction—show up with some trucks, and have them drive out of the area to the nearest town."

"Happy?" I said. "Happy for about a year, then what? Unemployment in a city most likely."

"What do you want from me, Sarann? Do you expect me to kick start a roaring economy in Bogota so that everybody can climb up the economic ladder until they can afford to line their home's floors with tropical hardwoods?"

"I'm sorry. It's complicated. Please continue," I said.

"We raze the farm so that the jungle can grow back. The roads become a one-way ratchet. You can leave on a road, but you cannot come back on one. The idea is to slowly depopulate these rural areas without causing a refugee problem. Once an area has been depopulated, we leave behind sensors to monitor visitors and move on to the next project."

"Where do you get the trucks?"

"We… borrow them."

"I see."

"We have assisted with the relocation of hundreds of families to date."

"What do you do when you run into resistance?"

"The farmers don't resist. The compensation they receive is substantial. They leave very happily."

"Do you have criteria for use of deadly force?"

"We do. Anyone firing at us, or caught laying mines in front of us, are at very high risk. Unarmed individuals may be handcuffed and detained awhile, but they are otherwise completely safe. We have video footage of most of our engagements. We may use it someday to defend our actions in court or possibly for public relations."

"What if someone just blows you off? I mean, what if you get some belligerent individual who knows you won't shoot him, and he refuses to cooperate?"

"Funny you should mention that. We had half a dozen journalists descend on us once. They ignored all of our warnings to leave. We couldn't shoot 'em, so we handcuffed them to a truck and sent them packing. They had no choice but to drive until they reached a town with a locksmith."

This didn't sound smart to me. "Weren't you concerned about bad press?"

"We had a real bit of luck there. We allowed one of the reporters to remain. He happened to be an old friend of mine. We gave him the royal treatment, including the ever-popular Sentinel suit tour. He was also given a bunch of footage showing our best boy scout behavior. He sees this movement in a positive light. The article he wrote was... "

I interrupted. "I think I read that article. Are you the one who subscribed me to that newsmagazine?"

"Maybe."

A little irritated, I continued. "That's wasn't exactly what I meant when I asked to be kept informed at the last think tank meeting."

"Hey, I've been busy."

"You can't possibly protect all of Amazonia this way. The job looks a bit overwhelming, wouldn't you agree?"

"Absolutely. We want to be the spark that starts a flame. We hope to someday see African and Siberian contracts with EPI. This needs to become a worldwide movement. Have you seen any of Jesús ' broadcasts yet?"

"Can't say that I have."

Bob continued. "I can sense that he's gaining momentum here in South America. You see his picture in public quite often now. I do not allow my people to carry anything affiliated with him because it is too soon to link him with our efforts. His support base isn't big enough… yet."

We left the farm and walked along in silence for a few hours, finally stopping for a rest beside another slow moving river.

"Your ride will arrive soon," Bob said. "It will transport you to a rendezvous where a car is waiting to take you back to town."

Waiting for the boat to arrive, I saw a dark shape silently materialize in the night sky. I continued watching as it quietly splashed down in the stream. Bob, along with two rangers ran down the bank to intercept whatever it was. Assuming it was safe, I also went down to have a better look. It turned out to be a twin-engined, two-man ultralight on floats. Painted jet black, it blended into the night.

"What do you call that?" I asked Bob.

"The bat mobile."

"Bat mobile—very fitting," I said. "How did you make it so quiet?"

"Have you ever heard of the Lockheed Skunk Works? It was a secret facility for special government military projects. In the 60's, they developed a super quiet spy plane that could fly over the jungles of Vietnam and spy on the Viet Cong. We just expanded on that technology. Our ultralight is a modified version of a standard commercial model. The engines are the key. They are rotary Wankels powered by hydrogen. They use some very effective noise suppression technology that makes them extremely quiet. The passenger can face aft and act as a tail gunner if necessary. It has heat sensor surveillance equipment on board which we use for nighttime reconnaissance."

Silence.

"So, when's the boat coming?" I warily asked.

"What boat are you referring to?" Bob said.

"Oh come on, you aren't expecting me to get on that thing are you?"

"Why not? It has two engines. It can hold altitude with just one of them. The pilot always follows waterways. If you were to lose both engines, he would simply do a power off landing in the river and request assistance. You have nothing to worry about. It has a very small radar signature, it's quiet as a mouse, and you will be flying out of range of small arms fire anyway. Trust me, they are a hell of a lot safer than helicopters."

Before I could protest further I found myself being strapped into the back seat.

"You have nothing to worry about," were the last words I heard Bob speak as we taxied out into the middle of the stream. The engines went to takeoff power and we were airborne in seconds. I closed my eyes and held on. We splashed down again an hour and a half later. I was soaked in cold sweat. We taxied to the riverbank where we were met by the same two guys who had delivered me to Bob. I changed back into street clothes and climbed into the car.

Two days later I was sitting in a stifling hot airport waiting to board my plane back to the states. I sorely missed my air-conditioned Sentinel suit.

This whole experience reminded me of Vietnam, only turned upside down. In Vietnam the firepower and technology of the United States was pitted against the stealth, efficiency, and dedication of North Vietnam's guerilla army. Bob has created a deadly hybrid, a high tech security force combining the best aspects of both armies. I wished I could think of a more innovative way to protect the Earth's remaining ecosystems. I hoped that this holding action would be short lived.

CHAPTER 7

The sermon

Bob had done a remarkable job of developing his nonprofit organization to protect ecological preserves. A huge stack of mail had accumulated while I had been visiting him and his EPI rangers in Colombia. I was digging through it. Bob had hinted that there might be another newsmagazine waiting for me, and sure enough, there it was. On the cover was a picture of Jesús standing in front of his cathedral in Mexico City. How Bob managed to get himself and then Jesús onto the cover of a national news rag like this is a mystery to me.

The story told of his startup ministry, his televangelist program, and his uncanny resemblance to popular renditions of Jesus of Nazareth.

Displayed in the cathedral's cavernous interior were several holy relics. One of particular interest was a tattered and faded blue rain jacket encased in glass. The story behind the jacket was going to be part of the sermon for his broadcast debut in the United States.

I put the magazine down and fumbled for my TV guide. Holy shit. It had already started. I tuned in to close-ups of young men and women singing time-honored choir music. The camera panned back revealing Jesús standing at the podium. His long jet-black hair coupled with a tall athletic frame contrasted sharply with the bizarre appearances that I have come to associate with most televangelists.

In a deep, compassionate voice that never fails to send a chill down my spine, he began, "The sermon for today is entitled 'The blue raincoat.' It is a true story. I know this with certainty because it deals with my own experiences.

"It begins with a camping trip in the mountains of northern Mexico. Accompanying me on the adventure was my six-year-old daughter, Carlita. It was early spring. Her uncle was also to accompany us but had declined at the last moment with other obligations.

"I had taken a similar trip to this place with my wife and daughter just two years earlier. An unexpected blizzard had materialized. We were ill prepared for such weather and were forced to move back down the mountain to safety. I had carried Carlita in my arms. My wife, Isabel had been following in our footsteps. At some point, I had realized that she was no longer behind me. By this time, Carlita had begun to shiver uncontrollably in the early throes of hypothermia; I had to continue down. Upon reaching safety, I alerted the authorities that Isabel was missing."

I detected a change in the tenor of Jesús' voice—restrained rage with an edge of grief. His sermon continued.

"The blizzard raged for two days. My wife, lost in the mountains, was never found.

"This camping trip was to be in her memory. Carlita had managed almost every step of the way. I carried her only the last mile or so. We set up camp and snuggled together reading books in our warm tent as night fell. Although the weather forecast was favorable, I was prepared for any kind of weather this time.

"Since her mother's death, Carlita had suffered from a sleep abnormality called night terrors.[47] Awakened by her screams, I would often find her cowering in a corner, staring into space, terrified by something only she could see. This could go on for hours. There is no way to comfort a child experiencing a night terror because she is not awake and there is no way to awaken her. She must face the nightmare alone. I was helpless, and could only wait until she closed her eyes again. Her physician had assured me that night terrors are most common when a young child is under stress as my Carlita had been. The child has no recollection of the nightmare the following morning and they say no harm is done.

"Tired from our long hike, we were soon fast asleep. I awoke a few hours before sunrise to the sound of a driving rain. I found to my horror that Carlita was not in the tent with me. I ran out-

---

[47] http://www.nightterrors.org/

side. It was a freezing rain; ice covered everything. Desperate, I searched for her and when I found nothing, I ran down the mountain for help. The local authorities rallied as fast as humanly possible and fanned out into the pouring rain and darkness. I was too exhausted to go back.

"Hours later, word arrived that she had been found—alive and well. I ran to her as they brought her down the mountainside wrapped in a faded blue raincoat; the sun was just rising. I cannot to describe to you....

Jesús paused, his voice gone. A few uncomfortable moments passed as he stood there in silence, struggling to regain his composure. Finally, he took drink of water and finished, "... my joy, as she put her arms around my neck."

"After the excitement of the rescue had diminished, I approached the farmer who had brought her down the mountain. He had found Carlita fast asleep under an old blue raincoat.

"I turned and saw the coat lying on a bench. I could not believe my eyes. It had belonged to my wife. She had been wearing it the last time I saw her alive."

Jesús paused, drew a breath, and continued, "Carlita had no recollection of what happened that night. I am confident that she had experienced a night terror and had simply walked out of the tent while asleep. How did this jacket, which I had not seen in two years, come to rest on top of my daughter?

"That was the day I met God."

I had to look away to get a tissue. I had no idea he had a daughter. During the commercial break I logged into the Internet and looked up some biographical data. My God... Carlita had died of complications while being treated for leukemia when she was seven years old, just one year after being rescued from the mountain.

The sermon continued with a touch of fire and brimstone. "Eight million children will not live past their first birthday this year. More than two billion people live in total abject poverty. Half of all childhood deaths are the result of malnutrition, each and every one of them held by a loving parent until the last moment.[48] Watching your child die is a nightmare without

---

[48] http://www.ncbi.nlm.nih.gov/entrez/query.fcgi?holding= npg&cmd= Retrieve&db=PubMed&list_uids=11100616&dopt=Abstract

comparison. What is God thinking as he looks at what is left of his green paradise? How does God view the extinction of one creature after another at the hands of his own children? God is not pleased. We have wrecked the Garden of Eden. Armageddon isn't coming, Armageddon is here."

The taped show continued for an hour. I don't normally make a habit of watching televangelist programming, but that was the most moving hour of it I have ever experienced. I hadn't seen Jesús in quite a while, and had a sudden urge to chat with him. I sat down in front of the computer and dialed him up. Much to my surprise, his image came up immediately.

"Greetings, Sarann," he said with a warm smile on his face. "It is good to see you again."

"Thank you," I replied, taken aback by my sudden success at contacting him. "Am I interrupting anything?"

"Of course," he said, the smile remaining. "However, I can always make time for you. What is on your mind?" He waved off somebody in the background. I could hear chairs shuffling, people talking, the sound of a door shutting, and, finally, silence.

"Did I just break up a meeting of some kind?" I said, feeling a little guilty.

"Indeed you did. Priorities are priorities. Please speak your mind."

I cleared my throat. "Well, I just finished watching your pre-recorded television sermon, and I have to tell you that I am deeply impressed."

"Thank you," he said. "I can only hope that others were equally impressed."

"Could I ask you some personal questions?"

"Please do," he replied, shifting his position slightly.

"Do you suspect that a large number of your viewers assumed that it was the hand of God that saved your daughter?"

"There is no evidence that God interferes in the lives of humans," he said. "Carlita found the jacket."

"Is that what your television audience thinks?" I asked.

He continued, "The idea that through the act of praying, God will help one football team defeat another, or even save a dying child, is irrational and degrading. What criteria would God use to intervene in such a random manner?"

It looked like Jesús wasn't going to answer that question. I moved on.

"It seems to me, Jesús, that if God does not interact with us, then isn't that essentially the same as saying that he does not exist?"

He smiled, "Sarann, we bestow honor and praise on people who demonstrate greatness. Although these people do not directly intervene in our lives, saving us from pain and suffering, we honor them for their greatness and for the joy that their greatness brings into our lives. There has never been and never will be a mortal that can compare to the greatness of God. Look around you at the splendors of this Earth. Isn't all life worth praising?"

I thought about this for a few moments, then asked, "Does the concept of Satan play a role in your religion?"

"The day you hold your dying daughter in your arms is the day you will meet Satan," he replied flatly.

"I'm sorry," I said and then clumsily changed the subject.

"I noticed that your ministry accepts three major credit cards and has a website. Are the charitable contributions coming in?"

"They are. I think this ministry has struck a nerve. People are responding generously. I may be able to wean my Church of your financial support soon. This will free up resources to do your other work."

"In your sermon you talked about God's children destroying the Earth on a never ending quest for status and wealth. Is that really how you see it?"

"Not exactly. I am in agreement with your point of view. People are not evil, greedy entities. Cutting down forests to build our homes or to make cooking fires is what we are supposed to do. There are just too many of us doing it. When a termite eats the foundation of your house, it isn't evil. It is just doing what it was made to do—eat wood. However, many people respond better if they perceive a struggle as good against evil."

"There are those who would call that just so much pop psychology hogwash," I said.

"Call it what you will. It was your head of state who coined the phrase 'Axis of Evil.' One of his predecessors borrowed the phrase 'Evil Empire' to describe the former Soviet Union. That word was carefully chosen in both instances. Explain to me why polls indicated that 80 percent of your fellow citizens approved of the attack on Iraq although they had no nuclear weapons program and had nothing to do with the Twin Towers disaster."

I know better than to argue with Jesús. "OK, you may have a point," I acquiesced.

Jesús continued, "Most children were brought into this world by accident; the result of carnal lust. My goal is the same as Bob's, to assist in the holding action that will slow the destruction of the planet's ecosystems while your contraceptive allows the people of the world to choose their family size with certainty."

"One more thing before I go," I said as I caught him looking at his watch. "I understand that your fan mail is primarily from women. Rumor has it that you have received a few marriage proposals. Do you ever plan to remarry?"

"I don't know, Sarann. I still grieve for my loved ones. I am not convinced that the pain will ever go away."

I thanked him for his time and signed off. Sometimes, I empathize too strongly with others. My sleep was troubled that night.

CHAPTER 8

Theme park

Following the progress of the think tank members had become a full time job. I had received an invitation to Dave's and Teresa's wedding and was on a flight to Jackson, Mississippi to partake in the festivities. Sitting next to me was a delightful elderly woman by the name of Edith.

"What brings you to Jackson?" she asked while we lunched.

"Two of my friends are getting married," I replied, sipping from a glass of white wine.

"Isn't that nice. What are they like… your friends?

"Well, Dave is an economist. He's seven foot tall, weighs 300 pounds and is afflicted with Tourette Syndrome," I managed to say through a mouthful of food.

"Oh really," she said, "… and the bride?"

"Um, Teresa is a biologist. She is a stunningly beautiful woman and a thalidomide victim, born without arms or legs. To say that they make quite a pair would be a gross understatement."

"Oh my. I see," she said as she reached up to push the call button.

When the flight attendant arrived, Edith said, "I think our friend here has had enough to drink."

A few hours later, I stepped off the plane and was greeted by Dave. He was beaming like the owner of a winning lotto ticket.

"Sarann!" he exclaimed with glee and grabbed me in a near fatal bear hug. "Come on, Teresa's out in the car."

We hurried over to a waiting Toyota Land Cruiser with heavily tinted windows. Dave threw my bags into the trunk. The car listed to one side as he climbed into the back. Much to my surprise, Teresa was sitting in the driver's seat.

"Welcome to Mississippi," She said. "Like my car?"

I looked around. The Land Cruiser had been retrofitted so that she could drive it.

"Nice. Very nice," I exclaimed trying to hide my shock.

I climbed into the passenger seat. She whipped the cruiser into traffic and careened across all four lanes like an Indy 500 racecar driver. I clung to the sides of my seat and was not feeling very well by the time we pulled into the park.

I was the first one out of the car. I needed air. Another car pulled up behind us. A man jumped out and walked over to Teresa's window.

Suffering from a serious case of road rage, he began, "You stupid bitch. Where did you learn to drive? Don't you know what a turn signal is for? If you weren't a woman I'd drag you out of that car and smack you sill… "

As he continued to rant, Dave opened his door. Watching Dave get out of a vehicle is like watching one of those clown acts where the clown manages to fold himself into a tiny little car, only in reverse.

As Dave stretched to his full height, the man stopped ranting, turned white, and began apologizing—profusely. He quickly excused himself, got back into his car, and drove off. Funniest thing I ever saw.

Dave walked around to the driver's door and scooped Teresa up.

"Where *did* you learn to drive?" I asked, dabbing beads of nervous perspiration from my forehead.

"It was no big deal. I've been tooling around in electric wheel chairs my whole life. As far as I'm concerned, this is just a three-ton, 240 horsepower wheelchair."

"I see," I said. "How do you manage to navigate off road?"

"Like this," she said as Dave effortlessly dropped her into a specially designed backpack.

We entered the park through a large decorative arch similar to those you find at places like Disneyland. It was crowded, with lots of tourists coming and going. The ticket takers waved us through. The walkways were paved and for the most part shaded. You could rent an electric golf cart if you wished. From each main walkway, side trails went off in every direction. These you had to walk because they were not wide enough for carts. There were signs along the trails warning to watch out for wildlife crossing. We came upon a small traffic jam. A crowd of tourists had stopped to watch a box turtle saunter across the path. A tourist picked it up and handed him around for the others to see. The children were very excited. A mother mallard followed by nine ducklings also crossed our path on our way to the first pond. In addition, we were escorted by a crow that insisted on sitting on Dave's head most of the way.

"What's with the crow?" I asked eyeing it with apprehension.

"There are about a dozen crows in this area that were raised by hand," Teresa replied. "They are very people friendly and the tourists love them. Our park staff raises lots of birds by hand—mallards, egrets, crows, and herons—to make them approachable. It's a gimmick to keep the paying customers happy. We don't do this with mammals because of the risk of bites and rabies. Being cold blooded, reptiles rarely transmit mammal diseases. Most of these ponds also have several friendly species of turtle and small alligators. These low-lifes are always on the lookout for food handouts. There is a stiff fine however, for anyone feeding any kind of mammal. Mammals that get too friendly are removed from the park, as are larger alligators and any poisonous snakes we run across. Waiver forms are signed by everybody entering the park as added insurance."

We arrived at the first pond. Waiting at the dock was a roomy air-conditioned swamp boat with large windows in the sides and bottom. As expected, several turtles and small alligators came alongside to beg. There were also several species of fish competing for elbowroom. Just for the hell of it, we threw them some food purchased at the gift shop.

Teresa continued, "Although the mammals are deliberately kept at a distance, many species have gotten used to people and go about their business ignoring the tourists. We've got it all, restaurants, hotels, swimming pools."

I was amazed by the quantity of wildlife to be seen. "How do you get wildlife to congregate like this?"

"We cheat," she said. "This part of the preserve has been grossly modified for tourists. It generates capital. We have provided birdhouses and nest sites for everything from owls to wood ducks. We have created a dozen grossly overstocked lakes and a maze of interconnecting waterways. You can pay by the hour to fish here."

It was probably my imagination, but the crow on Dave's head appeared to be staring at me.

Teresa continued. "We also run a farm that grows crops to feed everything from deer to quail to field mice. We have created an artificially enhanced ecosystem to generate a dense wildlife population. Populations like this naturally occur only around water holes during the dry season. We have made it a year round phenomenon here in the tourist areas. There's a lot for people to see. The rest of the preserve is off limits to practically everyone, no hunting, fishing, biologists, radio collars, darting, or tagging."

A dragonfly landed on my arm, and I realized that this should have been the height of mosquito season. "Where are all of the mosquitoes? I haven't seen one since my arrival."

"We control mosquitoes primarily with dragonflies," Teresa replied. "We have a dragonfly hatchery that churns out about a thousand a day. We also have a crew that is entirely devoted to mosquito control by organic means. As you can see, they do a great job."

Unlike other wildlife preserves where all you see are trees, everywhere I turned there was wildlife to be seen. This place had the feel of an open-air zoo without cages. Just then we came to the real zoo.

Teresa continued her monologue. "The zoo contains only animal species that can be found in this particular preserve. We use it for captive breeding and as a public education medium. The admission fees help to offset the cost of running the preserve. Roadkills, ecosystem degradation, pollution, hunting, and trapping are all partly compensated for by rearing wildlife to an age where it has a high probability of survival. Essentially, we are bypassing the naturally high attrition rate that is associated with newborn fauna. Most people don't realize it, but this part of the United States has the greatest diversity of vertebrate biology in the world. When you consider that roadkill accounts for about a million vertebrate deaths every day, Mother Nature could use all the help she can get."[49]

---

[49] http://www.sierraclub.org/scoop/bolgiano_roads.asp

We came to the turtle exhibit. It was a large cement pond with a glass wall on one side. Over thirty species of turtles were swimming around inside.[50]

"I had no idea there were so many kinds of turtles endemic to the southern United States." I said to Dave.

"Neither did I," he replied.

Not looking particularly cheerful, Teresa continued, "An illegal shipment of 7,500 Asian turtles was intercepted a few weeks ago in Hong Kong. Turtles are being caught and shipped to China by the millions.[51] China's growing economy may be the final death knell for a lot of endangered species."

Changing the subject, I asked, "Where did you find all of these enthusiastic employees?"

Dave answered, "We employ those unfortunate college students who made the mistake of majoring in fields like wildlife management or forestry. Typically, those who major in these fields either starve to death or end up going to law school. We have found the perfect niche for them. They're bright, educated, motivated, and about a dime a dozen."

While Dave was talking, his crow friend flew over and landed on my head. "Are you guys making any money on this operation?" I said, craning my eyeballs upward to discern the crow's orientation.

"We're not in the black yet. Our start-up expenses were stiff. The property was not cheap. Somehow, we need to get the word out that we are here."

That gave me an idea. "May I suggest that you contact Bob. I don't know how he does it, but he can probably get you on the cover of a major newsmagazine by next month."

"Are you serious?" Dave asked.

"Just give him a call," I said as I shooed the crow off my head.

"Would you be interested in seeing our security services?" Teresa asked as we walked past a brick building bristling with aerials and antenna dishes. "We hired two guys to handle this. They are usually holed up in here."

---

[50] http://www.ups.edu/biology/museum/NAherps.html

[51] http://www.wspa-americas.org/tv_story_turtle_rescue.htm

Inside, it was air-conditioned and comfortable. We entered a room where two men sat looking at computer terminals. They turned around to greet us.

Teresa introduced me, "Kleetus, Billy-Bob, I'd like you to meet Sarann."

I was taken aback. Between the two of them they didn't have as many teeth as I did. Local boys I guessed.

"Howdy," said Kleetus.

"Yeah, pleezed ta meet'cha," said Billy-Bob.

Teresa then directed my attention to a map on the wall. "The preserve is designed with several poacher traps around the periphery. These consist of easily defeated gates that lead to tempting looking access roads, which in turn lead nowhere. The sign at each entrance reads 'No trespassing please. This is a fish and wildlife preserve.' Each gate has a sensor that alerts us when it has been tampered with. We then turn on our surveillance cameras and track the trespassers into the preserve. The local law enforcement is also called to wait at the gate for when they return. Once the pickup truck has reached the dead end, a loud-speaker tells them to leave the premises. The police take it from there. It is very unusual for these people to leave their truck and six-pack to hike into the forest. For most of them, they need the truck to carry their beer. We have more cameras located in strategic places to help us track trespassers who enter on foot from places other than the more obvious traps. Apprehending them is just a matter of directing the law enforcement officials to meet them when they finally return to their trucks. The fines we manage to collect pay for most of the operation."

"Kids sneaking in with fishing poles are exempt," Dave interjected.

"That was Dave's idea," Teresa said and continued, "Our surveillance system is similar to that used in Las Vegas casinos. Our equipment, however, is much more complex. The casinos link everything with secure coaxial cable, and have a reliable power source. Ours has to deal with dirt, bugs, rain, and snow. In addition, we use cellular links and power sources that rely on solar panels to charge batteries. Kleetus and Billy-Bob designed the whole network using existing cellular telephone technology.

Kleetus has a PHD in electrical engineering and Billy has a masters in computer science."

"You surely cannot judge a book by its cover," Kleetus added with a sparsely toothed grin.

Suddenly an alarm went off. Kleetus jumped over to an unmanned terminal.

"Phone call from gate number 37," he said as he typed some commands into a keyboard.

A digital image appeared. It showed two guys in a pickup truck, nervously opening a gate to an access road somewhere in the preserve. The truck had a fully loaded gun rack. They started down the road with the camera turning to follow.

"There goes the first Budweiser," Billy-Bob gleefully exclaimed as a can flew out the passenger side window. Apparently, he and Kleetus had some kind of bet going. The local sheriff had been called and would soon be waiting at the gate. One camera after another came on line as the truck progressed.

"Whoa," Kleetus called out again. "Deer in the road."

The truck came to a stop, the passenger reached around for a rifle.

"We won't interfere if they intend to poach a deer," Teresa said. "The deer overpopulate anyway because this preserve has no grizzly bears, mountain lions or wolves to act as natural predators."

"The hunters have a heyday on the park boundaries come hunting season," Billy-Bob added.

A shot rang out, and the buck fell.

As we all stood around the console, Dave remarked, "Looks like these two have got themselves into a lot of trouble now. They'll have their guns and their truck impounded for this infraction."

They threw the buck into the back of the truck and headed for the gate. We followed their progress. As they stopped to let themselves out, two sheriffs appeared complete with the standard issue sunglasses and doughnut enhanced potbelly.

"I will be God damned! What in Jesus' name are you boys up to?" the older sheriff said to the surprised poachers.

The driver of the truck stammered, "Uncle Reemus, what are you doing here?"

"Boy, you best haul your sorry ass out of here before I tell your mom what you done. Bring that buck over to the barn after lunch. We'll have to dress it out before it goes bad."

With that said, the two boys hightailed it down the road, poached deer and all. There was a moment of silence and then Teresa said, "We still have a few bugs to work out."

As we left the surveillance hut, the crow settled on my head again. Teresa looked at me sideways. "He has really taken a liking to you."

We hiked along several more miles of trail and finally came around to where we started.

"So, let me sum this adventure up. You have managed to secure and return to the wild all of this river floodplain and marginal farmland. You are protecting this property from most human interference. At the same time you have taken a small part of the land and developed it into a kind of amusement park complete with hotels, swimming pools, fishing, hiking, and lots of wildlife to see and interact with. You are financing the management of the preserve with proceeds from the entertainment part of it. Have I missed anything?"

"That just about sums it up," Teresa said. "Your friend just crapped on your shirt," she added.

I reached up to brush the crow off my head, and he jumped onto my arm. I shook him off my arm just to have him land on my head again. Pausing briefly, I spotted a beanie hat at the concession stand, complete with a spinning pinwheel on top. I bought it and put it on. This solved the crow problem nicely, but stripped me of my last shred of dignity.

We returned to Jackson for the night. Tomorrow was going to be a big day for Dave and Teresa. Family and guests were flying in from all over. I retired to my hotel room to get some rest. Teresa's driving had worn me out.

The next morning dawned warm and sunny. I headed down to the cafe for a bite of breakfast. The waitress brought me the menu and a cup of coffee. As I mulled over my choices, I heard a familiar voice order a glass of juice. I turned around and saw an elderly gentleman sitting at the table behind me. He had a medium length white beard and glasses—a dead ringer for Colonel

Sanders. When he asked the waitress for more coffee I knew for sure.

"Bob?"

"Sarann!" he exclaimed, "I didn't know you were staying here."

Still stunned, I asked, "Do Dave and Teresa know you're here?"

"They do," he said. "I've asked them to keep it quiet."

I pondered this for a moment, looked around to make sure no one was in ear shot and asked, "What's with the disguise?"

"You don't want to know."

I *knew* that was coming.

"Won't you join me for breakfast?" he asked.

"Sure," I replied. "Don't tell me you're wearing a shoe phone."

"As a matter of fact, I am wearing a shoe phone," he replied, picking up on my reference to the old Maxwell Smart gag.

"I asked you not to tell me that," came my witty and well-timed response.

Both easily amused, we chuckled for a solid minute.

We joined up again later in the afternoon for the wedding ceremony.

You could not have asked for a more beautiful day to get married. The ushers had seated everyone in their proper places. A limousine pulled up; the door opened and out stepped Jesús.

"Doesn't anyone tell me anything around here?" I said to Bob, who was sitting next to me.

Jesús proceeded directly to the podium as the band struck up the wedding march.

The ceremony was not completely traditional. Teresa did not want to come down the aisle in a wheelchair. Instead, she and Dave came down the aisle together, Dave carrying her in the crook of his arm. It was all very moving. By the time the groom kissed his bride there wasn't a dry eye in the crowd.

"An absolutely flawless wedding," I said as I toasted the happy couple at the reception. Dave had been refraining from participating in any toasts. The excitement of the day was making it difficult to control his Tourette Syndrome. Finally, he decided he would give it a go. He stood up and tapped a spoon on the side of his wineglass. The crowd quieted.

I gritted my teeth.

"I would like to take a moment to express my gratitude. You don't know how much this means to both of us. Thank you."

The room burst into applause. Dave and Teresa's happiness was infectious that evening.

I turned to Bob and whispered. "He pulled it off."

"He sure did," Bob whispered back and added, "I'd sure like to be a fly on the wall at *their* honeymoon."

"You're sick."

"You have no idea."

CHAPTER 9

The Hyena

I managed to get myself invited back to Colombia. It had been six months since my last visit. There is nothing like real-life adventure, and I was looking forward to it. I was still flying with commercial carriers, but Bob had plans afoot for a personal long-range business jet. He hadn't decided yet between a business class 737, the Bombardier Challenger, a Lear, or a Mitsubishi Diamond jet. I was trusting in his judgment.

The flight had been uneventful. I braced myself for the equatorial heat as I exited the plane. I saw a friendly looking person holding a sign up that had my name on it. I walked over and introduced myself, "Hello, I'm Sarann."

"Pleased to meet you, my name is Carlos," he said, smiling, "I've been sent to retrieve you."

I was relieved to see that Bob had sent someone who spoke English and drove an air-conditioned Mercedes as well. There were two other rangers waiting in the car, a driver, and another in the back.

As I got into the rear of the car, Carlos introduced me, "Juan, this is Sarann,"

Without saying a word, Juan placed a gun to my head.

"Drive," he calmly said to the chauffeur.

Carlos pulled the door shut and the car sped away.

How could I have been so stupid? Incapacitating fear swept over me. I don't remember anything about the car ride. We parked in a run-down part of town. I was dragged into an abandoned building and up two flights of stairs. My assailants closed the door behind them, pulled ski masks over their faces and stood in silence, waiting.

It was stifling hot. A roach ran across the ceiling. Why the ski masks, and what were they waiting for? I heard a door open downstairs and the sound of footsteps. A well-dressed man flanked by four brutal looking bodyguards walked into the room.

"It is a great honor to finally meet you, Sarann," he said as he handed a video camera to one of the thugs.

He pulled out a butcher knife. "Get close-ups," he said to the bodyguard. "I want to record every last detail of this for our friend Bob. You know Sarann, I had the pleasure of doing this to another of Bob's associates. I believe his name was Jason. Do you know of him?"

I shook my head in the negative.

He continued. "Bob certainly does. He is a stubborn man and failed to respond in the manner I had anticipated. We will see how he responds this time."

I was in a state of shock. Everything seemed to happen in slow motion. As the man with the knife stepped forward, guns materialized in both the left and right hands of Juan and Carlos. Moving in perfect synchrony, laser sights blazing, all four guns fired as one.

The bodyguards slumped to the floor leaving the well-dressed man standing alone with four bright red dots jockeying for position on his forehead. The door behind him slowly opened.

It was Bob.

Still in a state of shock, I got up off the floor and walked past him. He turned to follow me out. As he pulled the door shut behind us the thunderclap of four handguns firing simultaneously shook me out of my stupor.

"That is the last time you fly commercial," he said, trying to instill some kind of gallows humor into the moment.

I was in no mood. "Explain to me what just happened," I demanded.

"All right. The cartel had sent two men to intercept you at the airport. They had bribed airport security to let them meet you on the tarmac, instead of in the terminal. I pay airport security a retainer fee which allows them to double dip. They get paid bribes and then are paid again when they report those bribes to me. We grabbed the cartel thugs sent to kidnap you, injected them with truth serum, and politely asked them where they had been planning to take you. I was not a happy Boy Scout when they said they were going to deliver you to the Hyena—a monster among monsters. I had rangers take their place and follow through with the plans. They put on ski masks so the Hyena wouldn't realize that a switch had been made. I could not pass up the opportunity to whack this psychopath."

"In other words, you used me for bait."

"Exactly."

Half disgusted, half amused, I shook my head while grinning with relief.

Bob continued, "You were not in as much danger as you thought. Juan and Carlos are two of my best men. They would have given their lives to protect you."

I looked around. There must have been a dozen rangers standing about, all in sentinel suits with full body armor.

"OK. I forgive you," I said. "Let's get out of here."

"Follow me," he said.

We walked out to the street and got into an old pickup truck. As we pulled away, I breathed a sigh of relief. My adventure had started with a bang, literally. Dusk was approaching. Cars were starting to turn their lights on.

"The U.S. military could take some lessons from you guys," I said, mulling over how smoothly my rescue had gone.

"That's true," Bob replied as he turned on the window wipers. A drizzle had started.

"Apparently, the U.S. military is keenly interested in our Sentinel suits. With some modifications, they would be ideal for Special Forces operations against real terrorists in mountainous or desert environments. They will have their versions of these suits soon enough. The only reason they don't have them already is because the technology making them possible has just matured. Being a government bureaucracy, they're slower on the uptake than we are.

"My greatest fear is that some politician or journalist will label us as terrorists. We have to avoid that at all costs. We do not want to be perceived as terrorists, rightly or wrongly. We only go where we have been invited. Officially, we are just a security service for nature preserves. This is purely a holding action until we can get a political ground swell, a South American green party with enough teeth to slow the devastation of the rainforests."

We drove on into the night. I was dog-tired. I hoped we didn't have much farther to go. The allure of sleeping in my comfortable Sentinel suit was not helping. I don't know how much time passed before Bob finally pulled over. We stepped out into the dark, frogs croaking all around us. A laser beam cut through the night.

"That's our signal," Bob said.

Standing at the edge of the jungle was a group of rangers. Bob and I quickly suited up. We were soon hiking through the night. Sunrise occurred a few hours later. We stopped to sleep. I found myself a nice place to hide and was out like a light.

I awoke at about two o'clock in the afternoon. A little early for an EPI ranger to be up and about, but I wasn't alone. Bob walked over.

"Better get something to eat. We'll be moving out shortly."

"I don't get it. I thought we only traveled at night?"

"We usually do, but today we are going to take advantage of some thunderstorms."

I could see how this made sense. We would be perfectly comfortable in our suits. Our noise attenuating systems would be able to eliminate most of the rain sounds letting us hear voices or vehicles through the storm. I ate quickly and was ready to roll. The expected weather hadn't arrived yet so I sat down quite comfortably in a nice mud hole to relax. One of my handlers was nearby. I decided to ask a few questions.

"Excuse me? Could I bother you with some questions?"

"Of course," came the always-polite reply.

"My name is uh, Pat by the way."

"I'm Scarface, pleased to meet you."

"How did you end up with Scarface for a nickname?"

He opened his visor.

"Whoa!" I said without thinking. "Well, you know what they say, 'chicks dig scars.'"

"I think there might be an upper limit to that hypothesis," he calmly replied."

"What is your real name?" I asked.

"Jason."

"Oh. I think the Hyena mentioned your name." Changing the subject, I asked him to explain how all of the night vision and laser beam stuff worked.

"I'll do my best," he said. It is pretty much the same technology the U.S. military uses. The lasers used for communication are in the infrared spectrum. They cannot be seen with the naked eye. We can see them only because our visors are designed to detect them. The beam must be aimed at the helmet of whomever you want to talk to. As you speak into your microphone, the sound is converted into a bunch of long and short laser pulses. The beam may look continuous, but it is actually pulsing very rapidly. Like Morse code. The receiving helmet converts those dots and dashes back into a voice that is played on speakers in the headset. The advantage of this kind of communication is that your enemy cannot intercept it. The laser can also be diffused and turned into a flashlight or spotlight that only those with the proper technology can see."

"I get it. But what if your enemy has technology to see in the infrared?"

"We use these lasers sparingly in any areas where that is a possibility. Our night vision equipment uses both image intensification and thermal detection devices. We choose one system or the other depending on conditions.

"The sensors in our helmets can amplify infrared light from the stars or the moon to visible levels. This light passes right through clouds so even on dark overcast nights we can see just fine.

"We can also see by temperature. This technology detects the different levels of heat given off by objects. All objects emit heat in the form of infrared radiation. Our sensors detect that heat by absorbing the infrared light, converting it into a video signal, and displaying it for us to see on our visors. This method works great in the rain because the cool air makes a good contrast with a warm body. Military and police forces use this same technology all over the world. It isn't anything new, really."

The rain had started to fall. I thanked my tutor and prepared to march. As the rain became a deluge, I activated my thermal sensing system and looked up to see a line of glowing shapes moving in front of me. I turned the system off and could barely see Jason. Night had fallen, and the monotony of another long hike settled in. I entertained myself for a time, shining my infrared helmet light on things and trying out the other systems. If the others found my actions annoying, they didn't let on.

An hour later battle language rippled through my headset. Our column came to a halt. I asked Jason what was up. "Landmines," he calmly replied.

A call came to me directly from Bob. Jason was to escort me to the head of the column to see something of interest. When we arrived, Bob and several other rangers were squatting in a semicircle around a segment of a road that our trail crossed.

"Take a look at this," Bob said as he motioned me over. "Turn on your thermal sensor."

As soon as I activated my system, I could see a number of gently glowing patches of dirt about the size of dinner plates.

Bob went on, "The mines absorbed heat from the sun during the day and are still warmer than the surrounding soil. We have deployed a radio controlled scout plane to see if the culprits are still within reach. Thousands of innocent people all around the world are killed every year and tens of thousands have their legs blown off by landmines. They should be outlawed. To this very day, people are killed in France from ordnance left over from World War I.[52] The French government has a special task force whose sole job is to collect and blow up old war ordnance, about six tons of it a year. The laying of mines in this conflict will not be tolerated."

More battle language over the headsets interrupted Bob's tirade. He then turned to me and began to brief me on all that was about to transpire. In return, I was to keep quiet for the duration.

"Our mine-laying friends are camped a couple miles down the road. A Dark Angel will be kept aloft to survey their activities and feed us live intelligence through the engagement..."

I interrupted. "What in God's name is a Dark Angel?"

He pointed straight up. I craned my neck to look and saw the dark silhouette of a flying saucer roughly fifteen feet in diameter, slowly passing overhead.

---

[52] Aftermath: The Remnants of War, by Donovan Webster

"Officially, it's called a Positional High Altitude Reconnaissance Tracker. For obvious reasons, we rarely refer to it by its official acronym, preferring, instead, to call it our Dark Angel. I…uh, wasn't thinking ahead when I named it."

"You didn't foresee that you would be asking your men to send a PHART aloft to watch enemy troop movements?" I said just to rub it in.

"Thought never crossed my mind," he said, and continued. "It's a hydrogen-filled balloon, shaped like a flying saucer, propelled by four silent, ducted fans driven by miniaturized versions of the hydrogen-powered Wankel engines used by the ultralights. The fans can swivel to provide thrust vectoring. It is remotely piloted, and will precede us in the night sky just above the jungle canopy looking for the infrared heat signatures of our antagonists. We will know the exact number and location of every participant before we engage. Each of our people has the option to project a map that shows the location of all participants onto his helmet visor."

I was stunned. The technology at Bob's disposal was unbelievable. "Wouldn't it be easy to shoot down a balloon filled with explosive hydrogen?"

"It would if we hadn't found a way to use hydrogen filled bubbles. The bubbles prevent oxygen from reaching the hydrogen and creating an explosive mixture. In addition, these bubbles are too big to fit through a bullet hole so the dirigible remains inflated. Pumping rounds into it will just make smaller bubbles out of big ones. Some hydrogen filled foam will slowly exude from the holes, but it will remain aloft. In fact, the covering is actually made of a fine mesh to save weight. It can lose three of its four engines and remain operational. It is not very fast or maneuverable and can only be used at night over jungle canopy, being too vulnerable to fly around in broad daylight. Stealth is its key to survival."

"What happens if it rains on a PHART?" I asked failing to contain my grin.

"It has a lightweight cover on the top—smart ass. One danger our rangers face as we approach the encampment are tripwire booby traps set up to secure the periphery. Our scouts are equipped with an ultrasound system that can detect trip wires.

The system searches the returning sound waves for any linear pattern fitting the profile of a wire. It then gives off a warning beep. Once the scout knows there is a trip wire a few feet in front of him he has no difficulty finding it. I got the idea for these sensors from the ultrasonic stud finders you can buy at a hardware store. Trip wires are being made obsolete by motion sensors anyway. We will avoid trails from here on just to be safe. Let's go."

A short while later we stopped.

"OK, we have moved Dark Angel into position above the enemy encampment at a fairly safe altitude—about 1500 feet. This rain will help. We will give our troops a little more time to get into their encircling positions."

As Bob made final preparations, I strolled over to where Oscar was standing. "Are you as scared as I am?" I asked.

"No. I have taken the appropriate dosage of combat medications. I would describe my mood as intensely alert, but not frightened."

"Figures," I said.

Bob walked back over to me. "All right now. Here is how it will play out. Dark Angel will drop a load of noisemakers and flash devices to wake up our friends. You will then hear our loud speakers coaxing them to surrender. We will tell them that if they choose to surrender, they will be disarmed, briefed, and then released. We call it our catch and release program—stole the idea from 'Trout Anglers International.' Three of the enemy in the encampment have already been captured by us once and carry our mark."

"How do you know that?" I asked incredulously.

"When we capture a cartel mercenary in the act of trying to kill us, he gets processed. First, he receives an injection that makes him...*comfortable*. Next, we tattoo a tiny bar code behind his right ear. Using a special syringe, we then insert a device that looks like a piece of multicolored fishing line into a vein in his scalp. It has a micro chip transmitter that allows us to identify him at a distance and if absolutely necessary, incapacitate him with an electric shock."

"A shock?"

"We stole *this* idea from the CIA, except their device is inserted into an artery in the brain and can be detonated. Our version sends them into an epileptic sleep seizure. One minute they are aiming a weapon, the next they are looking down a ranger's gun barrel. We should get pretty good participation here since there are former prisoners among them."

"What do you use to make them 'comfortable?'" I asked, knowing that he would run out of patience long before I ran out of questions.

"A combination of cocaine and Midazolam. We have an endless supply of the former."

"Isn't Midazolam an amnestic?" I asked.

"That's right." Bob said. "What do you know about amnestics?"

"Well, part of any anesthesiologist's arsenal is a class of drugs called amnestics. Your brain can't lay down new memories while these drugs are in your blood stream. In other words, you will not remember anything that happened while you were on these drugs. A surgeon friend of mine once told me that, in theory, making the patient unconscious or using painkillers is not necessary. All you really need to perform surgery is a drug to paralyze the patient so that he can't squirm around while he's being flayed open and a good amnestic so that he won't remember the experience and sue you. Your prisoners will get home and discover their tattoo, but have no idea how it got there. Same thing happens to drunken sailors all the time."

"What's the name of your surgeon friend?"

"Why do you ask?"

"I want to make sure he never operates on me."

"You're a real card today, Bob. What if you get some holdouts?"

"Oh, we're bound to get some hold-outs and some runners with a group this size. We will try several times to coax the holdouts to surrender, but in the end if they won't…"

"I get the picture. What's a runner?"

"They are the guys who make a break for it and head out into the jungle. Dark Angel and ground troops will track them down later if any get through our encircling forces. Remember, our guys can watch a runner on their map and move to intercept him."

"Got it," I replied quickly, sensing that Bob was nearing the end of his patience.

"We will split into two groups and bushwhack through the jungle on both sides of the road. When we get close enough, our own heat signatures will begin to show up on the transmitted display from Dark Angel. Our suits have transponders in them allowing us to instantly recognize each other. The enemy heat signatures will be a different color from our own.

"Here, put on this body armor. Stop! It goes on the outside of your suit. Jason can help you. He can also show you how to use this portable radar unit. It can detect metallic objects within a radius of about two hundred yards. Your target has to be carrying a gun or wearing a big gold chain to be seen. It has a much greater range than our infrared sensors. I want you to stay well out of harm's way. We will block the road behind and in front of our quarry and set up the usual prisoner corral, complete with bright lights and electrified fence to collect any volunteer non-combatants before we start firing. You stick with Jason here. Lie low. There will be a lot of stuff flying through the air. Got it?"

"Got it," I replied again.

Bob hustled off, leaving my bodyguard and me with our thoughts. Terse battle language crackled through the headsets. I watched as Jason swallowed his combat medications.

"What does it feel like to be on that stuff?" I asked.

"Like being on speed and Valium at the same time, hyper alert, but anxiety free."

"Interesting, Oscar pretty much said the same thing."

We moved off into the jungle. With Jason's help, I got the radar going and the armor on. By panning the small parabolic dish from side to side, I could see metal objects right through the jungle. The radar targets closest to me were a reddish color transitioning to blue for the targets that were the farthest away. I found that this dish could also be used to magnify sound coming from the direction it was pointed. I aimed it at the nearest radar target in the dark. Only the sounds of walking could be heard. Suddenly I bumped into Jason.

"Close enough," he said through the line of sight laser com. "My orders are to keep you out of harm's way."

"No arguments from me," I said.

We sat down and waited for the action to start. Transmissions from the Dark Angel dirigible high above the battle allowed us to watch the combatants' movements via the map projection on our visors. We were too far away to have our own heat signatures displayed. I watched intently as the engagement unfolded.

About an hour later, I heard loud speakers in the distance offering asylum from the coming firefight. I could see some dots on my visor screen moving to surrender. It seemed to be working. Suddenly there was a loud explosion.

"What was that?" I asked Jason.

"They probably walked into one of their own trip wires. Trip wires account for more 'friendly fire' casualties than any other weapon. That is why we never use them. Our electronic surveillance devices work much better. All hell is going to break loose now."

It did. The discussion was truncated by a series of extremely loud noises followed by flashes from a powerful strobe light. I ducked involuntarily. "...And that!?"

Jason calmly explained. "Dark Angel just deployed anti-night vision flares and ear splitters. The strobe flares emit an extremely bright light that wrecks the night vision of anyone who is not wearing eye protection. The ear splitters are bombs that convert all of their explosive energy into high amplitude sound waves that deafen anyone caught in their zone of effectiveness. The effects are not permanent, but they will last long enough. Our helmets automatically attenuate any noise above a certain decibel level and protect our eyes from bright flashes. Many of the enemy will be without two of their five senses for a while."

"I can imagine how hard it must be to fight if all you can do is taste, touch and smell your opponent," I said, trying to be funny.

Jason gave me the frequency to monitor more battle language. "Delta, Tango, Sierra...counter-fire radar showing three

AK's on semi off your 270, 295 and 350 radials respectively. Deadly force authorized...maintain radio silence."

"I think I know what he just said!" I was enthusiastically saying when a spray of bullets ripped through the jungle canopy. Another round smacked into a nearby tree trunk with enough force to rattle leaves off the branches. Jason motioned for me to lie flat. I was getting very nervous at this point—scared to death actually.

Jason's voice came through my headset. "Unfriendly radar target coming our direction. He's carrying a lot of metal and running fast."

I looked at my radar and saw the same blip. He was coming right for us. Suddenly he stopped dead...or did he? I turned my sound amplifier on. The sounds of a frightened and desperate human being running from certain death entered my helmet. He had thrown away his gun, which explained why the radar blip had stopped moving.

I aimed my COM laser at Jason. "He's unarmed!"

Jason moved into his path and waited.

Suddenly a bright floodlight clicked on from atop Jason's helmet directly into the eyes of the fleeing man. He stopped, the look of absolute terror on his face.

Sounding like Darth Vader, Jason's amplified voice boomed, "Tirate en el piso boca a bajo!"

The poor man did as he was told. Jason pounced, securing him with plastic restraints.

"Estaras a salvo, no tengas miedo." Jason said in a reassuring tone.

The three of us sat in the dark as the fight continued to rage.

An hour later, the sounds of the firefight subsided. The battle language over the headsets gradually converted back into a combination of Spanish and English. We were soon directed to a regrouping area. As the sun was rising, the sight that greeted us was grisly. Wounded combatants from both groups were receiving medical attention. I saw the shattered, bloody remains of a Sentinel suit helmet covered with flies. The taste of bile entered my mouth. I felt faint and sat down.

From my vantage point I watched the feverish activity. Dark Angel had been landed and disassembled. The parts were disseminated to several rangers for transport on their backs. In another area the injured combatants were being loaded back onto their trucks.

Bob walked over to me.

"I apologize, Sarann. This fight was not supposed to happen during your visit. I should have refrained, but the issue of landmines has to be driven home to these bastards.

"We need to get moving. It's daylight, the men are tired, and we are low on supplies."

I saw the fatigue in Bob's face. As our column started moving, a huge explosion from the direction of the road ripped through the air.

Bob anticipated my next inquiry. "We just blew up the landmines. That should have made one hell of a crater in the road. We are going to hike just long enough to put a safe distance between the road and us. We all need to get some rest. This would be a good time for me to answer any questions you have."

I felt that I was nearing the end of my physical endurance. Talking to Bob would be a good diversion. He probably already knew that.

"OK," I said. "Who were these people?"

Bob sighed. "The cartel decided to go on the offensive. They should not have laid those mines. That really pissed me off. We'll just cross our fingers that they'll know better next time. Next question."

"What do you do when you capture someone already wearing one of your tattoos?"

"Well, we've had some experience with that. We bring in one of our specially trained executioners."

My jaw dropped.

Bob threw his head back and laughed. "Don't faint. It's all an act. The poor guy with the tattoo is walked off to an isolated area believing he is about to be executed. You can imagine what that must feel like. Once out of sight, the *executioner* confides in him that he doesn't like killing people in cold blood. If he promises to run and never come back, his life will be spared. As he runs off, a few shots are fired over his head for good measure."

"I'm relieved to hear that, but what are you going to do if you catch a guy for the third time?"

"Fuck if I know. Give him a medal for bravery?"

That was one question too many. I could hardly stand by the time we stopped to rest.

I didn't even remember falling asleep. Jason was tapping on my visor to wake me up. It was pitch black outside. They must have let me sleep in. Everybody had obviously been up and about for some time. Jason already had a hot meal going for me. No sooner had I finished than we were on the move again.

This EPI ranger gig was getting rough. We came to a stream that we had to ford. As we crossed it, I thought about the other large predators of the South American jungles. The cayman is a relative of the alligator and crocodile. Less aggressive and slightly smaller, it is not usually a threat to a full-grown person. The giant anaconda water snake is another story. Sometimes reaching lengths exceeding your average house, these snakes can and sometimes do kill men. Then there is the electric eel. Hundreds of fishermen are electrocuted every year when they accidentally step on one of these bizarre creatures. Everybody knows about piranha. There is also a tiny fish with backward pointing spines that has the nasty habit of swimming up urinary tracts. I shuddered, the wrong thoughts to be thinking while crossing a muddy river at night.

Protecting a nature preserve in this part of the world would not be possible without the Sentinel suits. We would be eaten alive by leeches, flies, ants, mosquitoes, and other multitudes of biting, stinging insects; the most disgusting of which has to be the bot fly.[53] This strange insect hangs around waiting for a mosquito to come along. When a mosquito does show up, the bot fly grabs it, and they both fall to the ground. It then lays a few tiny eggs on the mosquito and releases it again. When the mosquito lands on someone, the sticky eggs detect the body heat and drop off onto the host. They quickly hatch into tiny grubs that bury themselves under the skin. A few days later a large red bump appears. Close inspection with a magnifying glass will show a tiny whisker in the center of the sore. That whisker is the breathing tube for a larva that has now grown as big as your finger tip and is happily living under your skin, eating your flesh.

---

[53] http://www.ambergriscaye.com/pages/town/botfly.html

A close second to the bot fly is the kissing bug, a member of a group of insects called assassin bugs. Bugs are subclasses of insects that don't have chewing mouthparts. Instead they have a pointy proboscis used to pierce and suck the fluids out of other insects or plants. The kissing bug sucks the blood out of people. It is called the kissing bug because it bites you on the lips while you're asleep leaving a red swollen sore. It also transmits a deadly disease. Few situations are more disconcerting than waking up to the sight of a bug the size of a man's thumb creeping up your pillow.

A flash of lightning interrupted my pleasant thoughts. Another tropical rainstorm had begun. I adjusted the temperature in my suit up a notch to chase away the chill. The group had stopped to rest and eat.

Bob walked over.

"We are going to pass by a land clearing operation on our way out of this area. They've hired armed guards, which suggests that they're aware that they are illegally logging on our property. We are going to neutralize their equipment."

"How are you going to do this?" I asked through a mouthful of mashed potatoes.

"We will hit them at night, as usual. Preferably we will siphon fuel from the equipment and use it to burn the engine compartments. This method is much more economical than blowing everything up. Our scouts have counted four armed guards. One of them has a heat sensitive night vision scope."

As I finished off my hot chocolate, I asked, "Doesn't that scope worry you?"

"Not really. They're useless against a fully cloaked ranger."

"Excuse me?" I said incredulously. "You have a cloaking device?"

Bob grinned. "Sort of. A suit in cloaking mode measures the ambient air temperature and adjusts the temperature of its own outer surface to match that of the ambient air. This renders the suit and its occupant invisible to heat sensitive night vision devices that work by detecting temperature differences. You've probably seen the police videos on TV where a helicopter with night vision capabilities watches the brightly glowing criminal try to shake his pursuers. In cloaking mode, an EPI ranger becomes the same temperature as the surrounding air and literally disappears from the night vision monitor."

Dumbfounded, I was temporarily without more questions. I finally came around. "Don't you wish there were some alternatives to this violence?"

"I surely do, Sarann. I wish that your TIFIC had existed fifty years ago when the population bomb first went off. Conflict is, however, an integral part of human nature. Some argue that it is a necessary condition. We are not fully human without it. When conflict cannot be resolved through peaceful means, violence is the last alternative. I know that doesn't sound politically correct. When you look at the causes of past violent conflicts, you'll not find a more worthy cause than this one; preserving as much of the Earth's ecosystem as possible for the sake of future generations. We aren't trying to wrest real estate away from others for our own use. We are simply saying 'off limits—go away—just pretend it is already gone and make do.' I am fully aware that every conflict has two sides, both completely sure that their cause is the right one. This conflict is no different from any other. Conflicts are not won or lost by who is right or wrong; they are simply won or lost. I do not want to go to my grave without at least trying to preserve the Earth for our children's children. The long-term future of humanity is at stake here. I know that many people argue that humankind can get along just fine without nature, whatever the hell that means. I can't buy that. Destroying Mother Nature is a risk that's just too big to take."

"Bob, do you know how pissed off I would be if I made a living raising cattle or logging trees, and some self-righteous bastard sneaked up one night and torched all of my very expensive equipment just because I was trespassing when I did it? Can't you empathize with these people?"

"I do. God damn it. You got a better idea? We do everything we can to dissuade them. When they return with armed mercenaries, we have no other option. I know these aren't bad people. They are trying to make a living. Lots and lots of lumberjacks have lost their way of life in North America simply because there are no more trees to cut down in their neck of the woods. Many states even have job-training programs that attempt to give these displaced families a new way to earn a living. It's an economic reality caused by a local lack of trees to cut down. We are imposing that same reality here. The only difference is that we are doing it *before* the rainforests are all gone."

Bob excused himself and moved off.

I trudged along thinking that I might have gone a little too far this time. I suppose if Bob gains enough worldwide support and successfully ropes off vast chunks of the planet until our population explosion ends, he will be seen as a savior. If on the other hand world opinion goes against him, he will be perceived as just another terrorist on the run.

Several hours later we came to a halt. We were near the logging operation Bob had mentioned. I waited with my bodyguard for the action to start. A surprisingly short time thereafter we received marching orders to hook up with the rest of the group. I caught up with Bob, puzzled that I hadn't heard any gunfire.

"What happened?" I asked, slightly out of breath.

"They were drunk. We poured diesel fuel on their machinery while they were partying, lit it off and just walked away."

How anticlimactic, I thought. This was to be my last night with the group. My visit was being cut short because of the heightened aggressiveness of the cartels. Bob hoped that this was a last-gasp effort to break him and that it was a good sign. I hoped so too.

CHAPTER 10

Reproductive freedom

As expected, the Vatican had taken a strong stance against the TIFIC. There had never been anything like it before—a contraceptive with the capacity to shrink the world population. Unexpectedly, most of the forces already arrayed against abortion had chosen to side with the Vatican. This was difficult for me to comprehend. Logically, with the TIFIC's potential to put the abortion issue to rest, you would think that they'd have been supportive. Instead, they chose to redirect their lobbying efforts against the legalization and distribution of the TIFIC. Their mantra this time—"Be fruitful and multiply; fill the Earth and subdue it." Tubal ligations and vasectomies never created such a fervor. Being painful, risky, and expensive, they never have been what you would call popular. They are also not easily reversed. The TIFIC made a lot of people nervous.

Arguments against the contraceptive lost most of their potency once its use became widespread. Restricting its use in the U.S. would have had no measurable impact since it was readily available in most other parts of the world and black market sources would have met the demand anyway. In addition, the number of new single mothers requesting welfare assistance had dropped steadily since the day the TIFIC was made available. Bob's favorite "pissing into the wind" analogy seemed especially poignant here.

The debate was reminiscent of that surrounding RU-486 abortion pill and the morning after pill. The media often confuses the two. Also called Mifepristone, RU-486 blocks progesterone necessary for sustaining a pregnancy and causes the uterus to contract, initiating a miscarriage. The morning after pill is a form of emergency contraception that can be taken up to 72 hours after unprotected sex. It works by causing an early period, thus preventing a potentially fertilized egg from attaching to the uterus wall.[54]

With the introduction of these pills, the pro-life forces had been blind-sided for two reasons. First, it was impossible to boycott or picket *every* pharmacy in the world that carried the pill. In addition, pharmacies that refused to carry them saw their profits going to competitors who did. Secondly, a black market supply was readily available. The battle lines drawn around many abortion clinics were circumvented overnight, a flanking action driven by biomedical technology.

Much to my relief, the TIFIC was very popular. Young men who took the TIFIC capsule found that they had a competitive edge when it came to securing the attentions of young women. The older guys also took to it because it was a lot less expensive and painful than the traditional vasectomy, and, unlike a vasectomy, temporarily reversible. Urologists who used to make a living performing these operations have weathered major changes, much like the lumberjacks in areas of the Pacific Northwest who have no more trees to cut down—phallic analogy not intended.

The effects were stunning. Statistics indicating that unplanned pregnancies account for about half of all pregnancies were vigorously verified. In areas where use of the TIFIC approached 100 percent, unplanned pregnancies ceased to exist. Having a version of the contraceptive for both men and women greatly enhanced its effectiveness. My suspicions that abortions would become rare had been born out in these countries.

The TIFIC has also caused vast unanticipated changes in reproductive behavior. I visited a village in Guatemala for a few months to learn for myself what some of these changes were. The contraceptives were made available in this area by packaging them along with free baby care products. Random quantities were sent home with new parents so that it wouldn't be obvious if one of the pills just happened to disappear. In this manner, one could discreetly take the contraceptive thus avoiding criticism or retribution from the local religious authority or disapproving relatives.

---

[54] http://www.prochoicecalifornia.org/s05factsheets/

Large families are the norm here. Typically, a married woman will have five children by the time she is 25.[55] After that age, she may continue to have children at a slower pace until she dies, loses her spouse, or can no longer have children for one reason or another. This pattern is repeated all around the third world.

Confidence in the TIFIC was not instantaneous. The popularity increased dramatically once the few women who had taken the contraceptive were able to conceive at a time of their own choosing.

It had been available for two years at the time of my visit. The birthrate in that very short period had dropped 50 percent. The prevailing notion among researchers, and among the families themselves, had been that large families were preferred. What did the introduction of the TIFIC do to cause such a radical shift in reproductive behavior?

I interviewed some parents with smaller than average families to see if I could find out.

Surprisingly, in several cases, both parents denied having taken the contraceptive. It would require a urine test to find out who, if anyone, was telling the truth.

In other instances, the mothers confided to me that they had taken the contraceptive so that they could control when their next child would arrive. Before the TIFIC, many unintended pregnancies became earlier-than-intended children.[56] The smaller families were simply the result of not having gotten around to taking the antidote yet. Procrastination afflicts all peoples. The vast majority of these tentative plans to have more children will never materialize.

In other cases, the father had taken control of the family's reproduction. Some men had taken the contraceptive as a safeguard while carousing outside of the marriage.

One woman told me that she had mixed the antidote into her husband's food because he had refused to take it. Marital behavior has some common threads across the planet. There is an old

---

[55] http://www.studentsoftheworld.info/infopays/rank/fecondite2.html

[56]http://www.hsph.harvard.edu/Organizations/healthnet/SAsia/suchana/0702/ravindra_berer.html

saying here, "When a woman wants to get pregnant, she will, and there is nothing her husband can do about it." That saying probably exists in one form or another in virtually all cultures.

Random urine tests in these communities revealed the most surprising finding of all. Almost all the men had taken the contraceptive. This result was not reflected in the surveys, which had indicated that very few men had taken it. Clearly, they were trying to hide the fact. One of the biggest concerns with this contraceptive is its potential to manipulate the reproductive rights of poor women. The TIFIC's popularity with men and the easy availability of the antidote has defused that potential problem. Women need not take the contraceptive at all as long as their partners do.

This contraceptive is just what the world needed—a reliable, safe, inexpensive capsule that makes infertility the default mode instead of fertility.

The concern that people would try to use the contraceptive to control the numbers of their enemies has become a reality. It is relatively easy to place the contents of a capsule into someone's salad dressing for example. I for one, find it somewhat amusing that the Israelis have taken to spiking the food supplies of the Palestinians instead of shooting at them. Significant numbers of Israeli men are also finding themselves mysteriously sterilized, probably by Palestinian sympathizers who work in the kitchens of Tel Aviv's restaurants. Once people realize they have been nailed, the antidote makes things right again. The TIFIC hasn't changed human nature.

It is quickly becoming widespread even in the countries where it is not legal. In fact, the black market demand is presently so great in Brazil a single capsule can cost hundreds of dollars. This won't last for long because the supply from the black market will quickly catch up with demand.

It is uncommon for the antidote pills to be made illegal. The Catholic Church has a hard time justifying restrictions on a medication that does nothing to inhibit conception.

A quarter of Costa Rica's landmass is set aside for ecotourism.[57] Its growing population, however, and the resulting

---

[57] http://www.sitesalive.com/crl/public/01f/crlgoals.htm

pressure to farm and log the preserved land is a constant threat to the government's political stability. It was the first Central American government to actually promote the use of the contraceptive. Presently, both the contraceptive and the antidote are being provided free of charge to all of Costa Rica's citizens.

I had been concerned about the potential for the TIFIC to accelerate the spread of AIDS. I suffer from recurring nightmares as is fitting for any megalomaniac out to save the world. One of them involves millions of emaciated, fly-covered orphans slowly dying of massive multiple infections. Typically, I wake up from such a nightmare and breathe a sigh of relief that it isn't real—but not in this case.

Some critics of the TIFIC insisted that it would promote rampant sexual promiscuity. That fear has not been born out and is probably one reason why AIDS has not increased in areas where this contraceptive has been introduced. The TIFIC hasn't suddenly made women start behaving like men. They are still the regulators of sexual activity. An interesting side industry springing up from the introduction of the contraceptive is the testing for sterility. Apparently, the girls aren't taking the guys' word for it. They want to see the results of the urine test first.

The TIFIC has also had an impact on some laws in the countries where it is legal. These laws are designed to protect women from men who claimed sterility, but who proved to be fertile when an unexpected pregnancy occurred. A stiff jail term followed by lifetime child support is typical punishment. God only knows what kind of wild litigation will develop in the United States.

Another concern I have with the contraceptive is that it might be distributed in the last of the hunter-gatherer cultures. The Pygmies in the Congo, the San bushmen in the Kalahari, or the Yanomami Indians of the Amazon, could all disappear overnight. These cultures could not afford the low birthrates that would occur because their mortality rates are still naturally high. Misused, TIFIC has the potential to destroy marginal cultures, instigating cultural genocide in a few decades.

The doorbell interrupted my thoughts. Looking through the peep hole I saw a middle aged man wearing a dark suit and sunglasses. I opened the door.

"Greetings," he said in a thick Italian accent. "My name is Michelangelo Pincanti. I am a member of the Swiss Papal guard and an officer in the Vatican security forces. You may call me Mike. May I have a few moments of your time?"

I didn't like the sound of this, but I invited him in anyway.

"Would you like something to drink?" I asked as he settled onto my couch. Safety was not a concern. Bob had seen to that. My home was fitted with security devices right out of a science fiction novel. The remote control I was holding had nothing to do with the television. The small monitor sitting on my counter contained a live image of Mike with three different crosshairs targeted on various parts of his body. With the touch of a button I could shoot him with an assortment of munitions: a bean bag bullet, a rubber bullet, or a pepper spray ball.

He declined the drink and got straight to the point. "An acquaintance of yours by the name of Bob may have been involved in the murder of the Papal guard commandant, his wife and a subordinate several years ago. Do you know his whereabouts?"

I was stunned. I didn't know what to say, so I just sat there.

Mike must have taken my inaction as an act of guilt and pressed home his perceived advantage. "I need to find him, and I am willing to go to great lengths to obtain the information I need." With that said, he stood up and pulled a gun from his jacket pocket.

I panicked and pushed the red button on the remote. Two-inch thick slabs of bullet proof clear lexan, the kind you see in front of bank tellers, dropped from slots in the ceiling, instantly sealing me off from my surprised guest. I carefully aimed the crosshairs shown on my monitor and simultaneously fired a beanbag bullet into his solar plexus and a pepper spray paintball to his forehead. I then picked up the phone and called Bob.

"Is that you, Sarann?" he asked.

"Damn right it is," I spat at him. "I could use some answers right now."

"What's happened?" he asked, sounding alarmed.

I went on to explain the events that had just transpired.

There was a moment of silence, and then Bob asked, "Does this man call himself Michelangelo?"

"He does," I snapped back.

"Whew," Bob exclaimed. "We have nothing to worry about. Mike is a certified, first class idiot—a real life Inspector Clouseau. He carries a cigarette lighter shaped like a gun. How bad did you hurt him?"

"Not bad, I think. You're telling me that was just his cigarette lighter?"

"Afraid so. I'll send someone over right away. Look, Sarann, this guy is a hack private investigator. I can't explain why the Vatican hires him, but he is their go-to man when they need work done in the U.S.; whatever he told you is bullshit. The introduction of the TIFIC has the Vatican grasping at straws right now."

"What about the murder of the Papal guards?"

"That happened way back in 1998.[58] I certainly didn't have anything to do with it. Hey, I've got a joke for you. How do you make a nun pregnant?"

"*Sigh*… I don't know, how?"

"Dress her up as a choir boy."

Bob was raised Catholic and has some serious issues yet to be dealt with.

I find it ironic that Italy, a Catholic stronghold, has a birthrate below replacement level. I know damn well that this is not a result of celibacy. The Church has been forced to change its doctrines many times in the past. If life is discovered on another planet, the Church will have to make some major adjustments, but they will. During the Spanish Inquisition, heretics were given the choice of converting to Catholicism or being burned at the stake. For those who chose to convert, the reward was strangulation at the hands of the executioner before the fire was lit. Maybe, just maybe, the Vatican's resistance to birth control will disappear in the not too distant future. You can't deny it; the Church has made significant progress through the years. Call me an optimist, but I'm confident it has the potential for more change. It should start with those silly squid shaped hats and then move on to do something about those striped balloon pants worn by the Papal guards… and many of Dr. Suess's characters.

---

[58] http://edition.cnn.com/WORLD/europe/9805/05/vatican.murder.on/

CHAPTER 11

The game park contract

Bob had been grooming Leto, one of his field associates, to head operations for a possible security contract in Africa. The opportunity presented itself when an invitation was received to assist with a severe poaching problem in a game park. The hope was that someone would eventually invite EPI to protect preserves in the Congo. Roads were being built into its heart at an alarming rate, and the time for action was now.

Born and educated in Britain, Leto was an ardent conservationist. A more natural leader I've never met. His deep voice was custom made to command. His skin was black as night. I can imagine that in the distant past he would have been a great African king.

I had first met Leto while observing the EPI operations in Colombia, so I recognized him immediately as I entered the airport terminal in Nairobi.

"Greetings, Sarann."

"Greetings to you, Leto," I said and added, "this is my first time in Africa, you know."

"I'll try to make it a memorable visit for you. This way," he cordially replied.

I was escorted to a waiting Land Rover. The day was sunny and cool. Our destination was a game preserve four hours away. This anti-poaching exercise was also being used to give the modified Sentinel suits—the Savanna suits—a field test.

Africa is a harsh place. Sometimes everything about it seems nastier than the rest of the world. The insects, snakes, plants, and wildlife in general seem to have a vicious demeanor. I was planning to walk lightly while visiting this part of the world.

The acacia tree, common in the savanna, has thorns three inches long. It is the unofficial plant from hell.

Africa is home to a poisonous tree snake called the black mamba. This is the biggest and fastest poisonous snake in Africa; its venom is also the most deadly.[59] It grows up to fourteen feet long, and is the only snake in the world known to deliberately chase a man down—a defense mechanism unique to the mamba. Some argue that the snakes are only mistaking people for trees and that is why they appear to be chasing you. Right… I once read an account by a snake collector who received a black mamba in a shipping crate. When he lifted the lid, the snake came boiling out and chased him about the room. He finally subdued it. It was the most terrifying experience he had ever had in a lifetime of handling poisonous snakes. He no longer keeps mambas in his collection.

Africa is also home to the spitting cobra, unique because it spits venom into the eyes of any creature it sees as a threat, blinding them.

Let's not forget the hyenas. They have a bite that is literally, bone crushing. If the bite doesn't kill you, the resulting infection from their filthy saliva will. Although accomplished hunters, they specialize in eating carrion. There isn't a rotten piece of flesh a hyena won't gulp down. If I were to eat one ounce of what these creatures relish, I'd be dead of food poisoning within hours.

There is more to spotted hyenas than meets the eye. They look like dogs but are not related to them. They are more closely related to weasels. It was once thought that hyenas were hermaphroditic. Females have a very realistic looking, but false set of male genitalia, and the reason for it is anybody's guess. Females are also larger and more dominant in hyena society.

African parks are islands surrounded by millions of desperately poor people. Some of them are armed with the ubiquitous Russian AK-47 assault rifle. These rifles are kept out of sight for fear of having them stolen. Millions of dollars were spent recently to move several rhinos out of Africa to Australia, to protect them from extinction. It was once quite common to assign a guard for an individual rhino. These armed rangers camped out near their rhino to protect it from poachers—a

---

[59] http://www.liberian-connection.com/tlc_snakes.htm

symbiotic relationship between a rhino and his human protector. The idea did not prove effective because once the poachers showed up, both the protector and his rhino were killed.[60]

There are data to suggest that the disappearance of the megafauna that once roamed the Earth—the mammoth, giant ground sloth, and woolly rhino, to name a few—coincided closely with the appearance of man in their respective parts of the world. Why then, do African megafauna—the elephant, rhino, hippo, and giraffe—still exist? Some think they still exist because they co-evolved with man and developed adequate avoidance defenses against him. I think that is a reasonable hypothesis. There would be nothing left alive on the Galapagos Islands bigger than a finch if man had not *deliberately* spared much of the wildlife. On the other hand, the American bison managed to dodge extinction at the hands of primitive man. It was finally driven to the brink of extinction with industrial age technology, and only saved at the last minute. I believe the megafauna of Africa would have been driven to extinction at the turn of the century if it were not for game parks. Someplace had to be the last to harbor megafauna, and that place is Africa. Intervention by concerned conservationists has temporarily saved them and that is the only reason they still exist. In all likelihood, it was the combination of a rapidly changing environment—the end of an ice age—as well as the introduction of man that destroyed the megafauna on the rest of the planet—a one-two punch.

Leto and I arrived at the game park in the late afternoon. We were greeted by the staff and invited into the lodge for drinks and conversation. Our hosts had insisted that we stay at the lodge instead of camping on the cold hard ground. We declined. The Sentinel suits were perfectly suited for the Congo. A modified version was to be used for the drier savanna. Our hosts had no idea that an air-conditioned Savanna suit would be far more comfortable than anything they could offer.

Leto had already established a camp several weeks before my arrival. The EPI ranger staff was allowed minimal contact with locals. This was partly to prevent excessive curiosity about the Savanna suits. When we arrived, I could see nothing. It can be very dark in the countryside away from city lights. An EPI

---

[60] http://wildnetafrica.co.za/wildlifenews/2000/03/116.html

ranger camp is hard enough to spot in the daytime. It is invisible at night. A group appeared out of nowhere to greet us. They had our suits ready. I was becoming quite proficient at putting these on. At some point, I had to be briefed on the differences between the Savanna suit and the Sentinel suit. This would be the job of my assigned bodyguard, Duncan. The briefing could wait until morning. I was dog-tired and once my suit was on, I was ready to find a discreet place to just lie down and crash. I adjusted my night vision system and found a cozy spot well within the camp periphery to sleep. I had to keep reminding myself that this was Africa. Sleeping on the ground here could be hazardous to one's health. Before drifting off in air-conditioned comfort, I looked up just in time to see a falling star. How cliché.

Morning came fast. I sat up and wiped the dust from my visor. Dust is the nemesis of mechanical equipment everywhere. It would be interesting to see how well the suits performed in this environment. As expected, Duncan was close by.

"So, Duncan, what's different about these suits?"

"I may as well brief you while we prepare breakfast. As you can see the Savanna suits have a different color and camouflage pattern. They blend into this countryside extremely well. The major concern here is dehydration. Sources of water are scarce. These suits, when kept sealed, condense water vapor from your breath by running it over the cooling coils. Condensed vapor and urine are all collected in a reservoir located on your left leg. Someone in each patrol is assigned to carry the water-recycling unit. Once a day everybody brings their wastewater to him. The purified water is redistributed later. We can greatly extend our patrol range during the dry season by using this system."

I interrupted, "Somehow, this all sounds too familiar—a suit that recycles water. Never mind. It will probably come to me later. Please continue."

"Our fuel cells do not last as long in this environment because of the heavy power drain from the suit's air-conditioning system. We can only go for five days between recharges. Let's see, have I missed anything? Oh, yes. The pills in this pouch are to be taken at first sign of anthrax illness. The dry soil of the African savanna is rich in the spores."

"Why is that?" I asked.

"Although microorganisms often kill their host, they did not actually evolve to do so. The death of a host is usually an unfortunate by-product of too many organisms using the host's resources in a non-sustainable manner—sound familiar? When the host dies, so does the colony of infectious organisms. It becomes a lose-lose situation. Anthrax is highly unusual in that it has evolved to kill its host. That is how it propagates the next generation. The infected animal ends up dead on the savanna, and scavengers tear the carcass apart. The infected blood seeps into the soil where the bacteria form a durable spore that can lie dormant for years until another herbivore whiffs it up its nose while grazing and repeats the cycle."

"I guess I'll just add anthrax to the list of nasty African organisms to watch out for during my visit," I said with a shiver. "What kind of tactics are planned for this operation?"

"First, we are operating in the open savanna. Second, our means of getting about is radically different. We use the two man ultralights you saw in South America almost exclusively. They are quiet and can move many times faster than ground transportation. We use helicopters only for occasional re-supply.

"Once a band of poachers gets loose inside the park boundary and into the bush country, you have a needle in a haystack. That is why they must be detected as they penetrate the periphery. The odds of finding them go down rapidly if we are not successful within the first few hours after they get inside."

"Didn't you have the same problem in Colombia?" I asked.

"The terrain was different there. Here, a person can walk in any direction. In thick jungle, you are forced to use trails, roads, or rivers. Monitoring roads here is pointless.

"We have put some new technology to work on the problem. Like most African game parks, this one is encircled with a road and a fence. The fence acts as a barrier to protect the wildlife from the teeming masses of humanity, and it discourages wildlife from moving out of the park boundaries and becoming part of the bushmeat industry. People, however, easily defeat fences. We have just finished installing a new system."

Duncan pulled a small device out of his pocket and handed it to me. It was a camouflaged box with an antenna, a camera lens, and a small solar panel.

"We have installed thousands of these around the park perimeter, just inside the fence. Each is a combination infrared motion sensor, digital camera, and cell phone. The precise global position of each unit is recorded so we can find them if they fail or need maintenance. A computer constantly monitors them. The system detects dozens of false alarms every hour. It filters most of them out based on the shape of the infrared heat signature. A human has a unique height-to-width profile. If the computer decides that the heat signature might belong to a human, a digital picture is taken and transmitted to security. Security then scrutinizes the image and decides whether to send a patrol or not."

"How do you find the poachers once they have passed the ring of detectors?" I asked.

"For that problem, we rely on a combination of new technology and old—GPS transmitters on specially trained tracker dogs. When these dogs detect a scent, they are released to follow it. Once they locate their quarry, they hold perfectly still while pointing their noses in the direction of the trespassers, like a bird dog would do.

"The new technology comes into play with the global positioning transmitter around the dog's neck. As soon as a dog goes into pointing mode, its location can easily be discerned. This same technology is used on hunting dogs in the United States and Europe. If the dog loses the scent or needs to be called back, an old-fashioned dog whistle suffices.

"The major problem with any kind of security organization is that most of the time there is nothing going on. People will always let their guard down. Vigilance slides into routine, routine into boredom, and the result being that an intrusion fails to be detected.

"Leto has developed a means of dodging that problem. The security teams are organized as competing groups. They take turns simulating poachers. When a dog team finds a scent trail, they never know if it is for real. It doesn't matter. Tracking whoever it is presents a challenge and excitement. A score is kept, and trophies are awarded periodically to the most proficient teams—one for detection, one for evasion.

"We have three distinct goals to assist the game park. The first is to provide the firepower to get the poaching problem under control. The second is to equip the rangers with the necessary security technology. Lastly, we will train the park staff to implement intrusion detection and suppression."

"So, Duncan, are you protecting this preserve from people who have traditionally hunted these lands or are you protecting it from people who want to profit from rhino horns?"

"What is the difference?" He replied.

I had to think about that one. Killing game for meat or profit is a very fine line. The local's ancestors had undoubtedly driven out some preceding group of people to gain control over this land's resources in the first place. Noble savages, my butt.[61] I guess this is just the latest in a long series of hostile takeovers.

"Well, it looks to me like you have things under control," I said. "Mind if I poke around until something comes up?"

"No problem, just keep your COM mike open."

I've always wanted to explore Africa. Most tourists are interested in the big game animals. I am interested in the smaller stuff. I hiked over to a rocky outcrop and looked around. Turning over rocks is a great way to find wildlife. It didn't take long. I found an emperor scorpion about the size of my foot under the first stone.

Eventually I stumbled onto a fresh burrow. I wondered what was down there—maybe an aardvark or a warthog. I decided to sit quietly and watch. My patience was finally rewarded. A whiskered nose appeared first, followed by the biggest rat I have ever seen. It was a giant African rat.[62] I had read about them but I had never seen a live one. It looked exactly like any ordinary rat except that it was the size of a cat. Imagine New York City's sewer system filled with these babies.

Leto's voice came through my headset, "We have detected trespassers in our sector, hurry back."

Upon my arrival, I saw that five twin-engined ultralights had materialized out of nowhere. Their engines were already running, but I hadn't heard a thing until I was within about 20 feet.

Leto's voice came through my headset again. "Your ride is waiting."

---

[61] Constant Battles by Steven A. LeBlanc with Katherine E Register

[62] http://www.rmca.org/Articles/giant.htm

I saw him waving from the only ultralight without a second passenger.

"Oh, shit," I said aloud.

As I clambered into the back seat, he said "No time to brief you, just put your seat belt on and sit tight."

I did as I was told as all five craft began their takeoff rolls. We were airborne in seconds. Once aloft, we spread out into a loose formation. As I looked at the other ultralights, I realized that two of the passengers were dogs.

We flew on this way for about an hour. It was quite pleasant once the shock wore off. Unlike riding in a typical ultralight, with the scream of the engine and the spinning prop inches from your head, the propellers in this one had been replaced with fan blades encased in circular shrouds similar to those at the front end of a modern fan-jet. Our Savanna suits were perfect for this kind of flying. In the days of open cockpit planes, the pilot and passenger wore goggles, scarves, leather jackets and hats to keep from freezing to death. I was quite comfortable.

I opened my visor to see what the wind felt like on my skin—big mistake. I was picking bugs out of my teeth the rest of the day. I saw Leto pull a rag out to wipe his visor off. I reached into the same pocket on my suit and imitated him. The rag was soaked with a bug cleaner of some kind. They think of everything, these EPI rangers.

Two ultralights landed, and the rangers fanned out to provide a secure perimeter. The others circled, ready to provide cover from the air. Finally, the remaining ultralights landed one at a time. They were then partially disassembled and hidden.

"What do we do now?" I asked, feeling out of place with no duties to perform.

"You stay with me for safe keeping," Leto replied.

The rangers, spaced about a hundred yards apart, formed a line. The dogs quickly picked up the scent and were released into the bush. Leto and I stayed behind with the ultralights.

"So, what's the game plan?" I asked.

"If we find the poachers, we will shadow them until nightfall. If we don't find them today, we will send an ultralight aloft after dark. We should be able to find them with our heat sensing equipment, especially if they light a fire."

"How do you land an ultralight in the pitch dark?" I asked.

"With our night vision technology it's doable," he replied.

"Doable?" I said.

We waited. I found myself a shady spot under an acacia tree and lay down to read an e-book on my visor display. Before long, my communication set crackled to life. It was the patrol reporting in—no luck. It was up to us now. They would camp in the bush until we gave them the coordinates of the poachers. As dusk descended, we readied the ultralight. I realized that I would be of little value as a passenger and this flight was clearly not going to be a low risk venture.

"What do you wish to do?" Leto's voice snapped over the radio. "You may wait here in the dark until I return or you can join me… your choice."

"I'm in. Let's go." I replied. Sitting alone in the African bush after dark just didn't appeal to me.

I climbed on board, and we were off into the African night. As we gained altitude Leto filled me in.

"We will go to about 3,000 feet. We have the advantage of knowing where they aren't, thanks to the patrol today. We will execute a grid pattern using global positions punched into the autopilot computer. Keep your eyes peeled for a campfire. I will concentrate on the infrared signatures."

About twenty minutes into our flight I saw their campfires off to our right. Leto saw them also and immediately veered off to avoid being detected. We passed the global positions on to the ground forces and returned to our point of origin.

As Leto circled to land, I tried to see where he planned to set down. The night vision was grainy, not at all clear. I decided to close my eyes and just hold on. A couple of jolts later, we were on the ground again.

"You've got guts, Sarann. Landing one of these at night is frightening enough, but to be the passenger with no control whatsoever must be especially unpleasant."

"No big deal," I squeaked.

With the ultralight secured, we headed off into the night. Several hours later we rendezvoused with the others. We now knew precisely where the poachers were relative to our position.

The group carefully deployed to surround the encampment and execute the battle plan. There were eight in the poacher group—all of them armed with AK-47s. They would know how to use them.

"How do you arrest eight heavily-armed poachers?" I whispered.

"You don't," he calmly replied. "Poachers are shot on sight. It has been the policy in most African game parks for decades."

"That's sick." I protested. "You can't shoot people who are just trying to make a few dollars."

"I didn't make the policy. Attempting to arrest them will get some of our own killed. That is a fact, learned the hard way. They know the rules as well as we do."

I continued. "That was before you guys showed up. Poaching has become a suicide mission. This isn't right."

The rangers continued to set up their weapons.

"You can't do this, Leto."

"Very well. Duncan, deploy a strobe light at these coordinates along with a loud speaker."

"Thank you," I said.

A short time later the light was switched on and the speaker blared, "Drop your weapons and walk toward the light!"

Had the poachers known that they were completely encircled by EPI rangers equipped with night vision technology, their response might have been different. There were no takers. They began firing toward the light and the speaker.

I flew back in the medevac chopper with the one survivor. I'd seen more violence than I cared for.

A few days later Leto asked, "If you are feeling better, would you be interested in watching a disarmament mission?"

This was another idea of Leto's for bringing the poaching problem under control—disarm the young thugs in the villages that surround the park.

"Not if it involves killing people," I said.

"We will do all that we can. What do you say?"

"O.K. I guess. Can you at least give me an idea of what this operation will entail?"

"Surely. We start with a weapons buy-back program. That is usually all that is necessary. While we are in the village, we also pay a bounty for information leading to hidden weapons. This also works quite well. We finally fly a dark angel over the area in the middle of the night using air-to-ground radar that can detect metallic objects buried under three feet of dirt."

"You know, Leto, those weapons belong to those individuals. You're stripping them of their most valued and prized possession."

"Pathetic, isn't it," he replied "that their most valued possession is an assault rifle designed to kill people? The Masai's most valued possessions are the spears used to kill lions to prove their manhood. These guns, however, are used to rob people and poach game."

He continued, "If we detect a significant cache of weapons, we move our forces into place under cover of darkness. The huts that hide the individuals with weapons are targeted in the wee hours. Then we use Bob's tried and true method of shock therapy—a combination of deafening noise, blinding flashes of light, darkness, and overwhelming force. You can't disarm just a few villagers. You have to disarm all of them. It only takes one thug with an AK-47 to bully everybody if he's the only one left with a weapon. Come on, let's roll."

With that said, I followed him to the ultralight. The others were preparing for the flight as we arrived. As soon as my harness was fastened, Leto yelled "Hold on!"

The adrenaline rush I get while straddling one of these frail flying machines is almost too much to handle. It scares me half to death. Fortunately, with the Savanna suit I didn't have to worry about peeing myself.

The sun set just as our flight crossed the park boundary. The landings were uneventful.

Leto and I took up a position where we could watch the operation from a safe distance with our night vision scopes.

The next thing I heard was a string of deafening explosions. They made my ears ring even from our vantage point about a quarter mile away.

Intermingled with the noise was the usual series of intensely bright strobe flashes intended to cause night blindness. The surprised villagers put up no resistance. Each hut was quickly searched and all lethal weaponry removed.

"That was fast," I said with relief.

"As designed," Leto said. "Would you like a ride back to the ranger camp?"

"No thanks." I said. "I'll go in the Land Rover in the morning. I don't think I could handle another night flight right now."

Later that night we set up a camp perimeter. While looking for a level spot to bed down for the night, I came upon several old warthog dens. These abandoned holes are a favorite hiding place for the African rock python. I once saw a photograph of a 25-foot long python with three zookeepers sitting along its length. The luckiest keeper got to sit just behind the snake's shoebox-sized head and was grasping it by the neck with both hands. These snakes routinely eat antelope, horns and all, giant African porcupines, quills included, and even the occasional full-grown wart hog or leopard.

The next morning I rode back sitting atop a pile of confiscated weapons and ammunition. Leto was waiting when we arrived, "I see you survived the night. Did you sleep well?"

"Not really," I said. Having dreamt about snakes all night, I did not feel like explaining so I changed the subject. "I was wondering how you trained your men to such a high level of proficiency?"

"Some of them are veterans of the South American job, but their training for operations in villages and towns came from paintball games inside warehouses."

"Seriously?" I inquired. "How does that work? "

"One can find indoor paintball arenas all around America. Typically, they are large plywood castles complete with dungeons, ramparts, and turrets constructed inside a big warehouse. The operators rent guns and goggles to groups who reserve time to play.

"The first thing new recruits learn is how easy it is to be shot. Rambo is the only person on Earth who can stand out in the open and mow down an army without being scratched. In a paintball firefight, most participants are shot in the first five minutes. One quickly realizes how hard it is to stay alive. The paintball that gets you comes from someone you never even see. That is why our weapons have all been designed to look and shoot around corners without exposing the combatant.

"Once our guys were trained and equipped with paintball versions of our equipment, we became invincible. The best paint-ball teams in the world were like putty in our hands."

"It makes me wonder why our military never came up with these weapons themselves," I said. "How can you people outdo the U.S. military in R&D?"

Leto laughed. "We have not outdone them. We have simply taken a different tack. They spend billions of dollars a year on R&D. You don't seem to have an accurate image of the military industrial complex. It is big, and it is fat, that's for sure. However, it is anything but innovative. Most people do not realize that the military hasn't the capacity to engineer or build anything. Everything they have was dreamed up, designed, tested, and manufactured by private enterprise. The top executives, or generals as they prefer to be called, are some of the most unimaginative, politicized bureaucrats you will find anywhere. Now that they have seen our technology, they will soon have copies for their own use, built for them by private companies as always. The Viet Cong bled your military to a standstill using tunnels and the Ho Chi Minh trail. They shot down well over 4,500 helicopters, each worth millions of dollars, using a few thousand dollars worth of small arms munitions. Please, don't get me started."

Man, did I hit a nerve.

The next day I was planning to watch my relief organization's introduction of the TIFIC to a local village. I was keenly interested in how this would go.

I arrived along with my bodyguard early in the morning, ahead of the agency's crew. We would be perfectly safe. An EPI ranger enclosed in a Sentinel suit, visor down, and bristling with antennas, is a terrifying sight to the uninitiated.

I hung around the village. Once the inhabitants got used to me with my visor up, I was able to talk to the women and even play with the children. The men kept their distance. Finally two trucks appeared. They had brought along a tent, visual aids, the TIFIC, and an interpreter.

They set the tent up and invited everyone to come see the new medicine. The spiel went something like this: "If you take this medicine, a man cannot make you pregnant against your will. You can only get pregnant when you want to. To become pregnant, you must take the blue pills before okwebakaka—having sex."[63] A similar story was put together for the men. Mixed in with this was a lot of material about safe sex and HIV.

Not everyone fully understood the concept. Some of the villagers were convinced that the blue antidote pills would make them pregnant.

Through all of the confusion, I was frankly surprised to see so many villagers take the TIFIC.

Large quantities of the antidote pills were left behind in sealed packages designed to be weather resistant. They had also been modified to be overdose tolerant to prevent children from accidentally poisoning themselves should they get into the supply. Access to the antidote must be given a high priority if people are to trust and use the contraceptive. Time will tell.

On my last day at the game park, the rangers had a farewell feast. We dined on roast wildebeest, the result of a roadkill by a ranger's Jeep. People are a hazard to the other animals on the planet even when we are trying to protect them.

---

[63] http://www.google.com/search?q=okwebakaka&sourceid=mozilla-search

CHAPTER 12

Indiana refuge

Teresa and Dave were at it again. This time they were grabbing a chunk of Indiana. The northern end of the state is flat as a pancake and resembles a giant cornfield. However, as you move south, it begins to get hilly. The hills become more pronounced as you move into Kentucky and finally become the Blue Mountains of Tennessee.

I was born and raised in Indiana. Tan and corn-fed, growing up healthy and strong on the family farm did not describe my childhood. I grew up in a ghetto on the south side of Indianapolis. A single road separated the black neighborhood from the white one.

Even in this slum on the edge of a big dirty city, there were the weed lots, the places where kids could go to see nature. We called one of these weed lots "the swamp."

It truly was the remnant of a swamp. Every spring, large black salamanders appeared under the ice to spawn. Great horned owls could be seen in the trees. It was a child's dream playground, a few acres of knee-deep swampland. My older brothers would come home with voles, shrews, and anything else they could catch, huge praying mantises sitting on their shoulders like parrots from another planet. As is the fate of virtually all childhood memories of the corner weed lot, it is gone now, paved over.

My uncle Clyde had a farm of sorts. Perched on a high hill in southern Indiana and surrounded by forests, it was the source of my fondest childhood memories. Visiting the farm was always exciting. Living on an isolated farm however, can be mind numbing. My cousins were always very happy to see us.

I remember how Box turtles would appear out of nowhere after thunderstorms looking for earthworms to eat. There were always pet box turtles running around in the house. We had to wear socks at the breakfast table to keep them from nipping our toes. I recall the eight-foot-long black rat snakes climbing in the apple trees, the king snakes in the cellar, and the copperheads in the barn. A walk around the pond would send dozens of frogs into the water. The big farm light on the telephone pole stayed on all night. The light attracted insects, which in turn attracted bats and toads. Cecropia and Luna moths the size of a man's hand were common.

I remember the cicada killers, solitary hornets two inches in length, with sinister looking, hooked black stingers. They patrolled the edges of the woods looking for their cicada prey. You always knew when they found one because the cicadas would make a God-awful racket as they were being dragged back to the hornet's nest. Once there, the hornet stings the cicada into a state of paralysis. The hornet then lays its eggs and plugs the tunnel up. Upon hatching, the hornet larvae proceed to eat the cicada alive, saving the critical organs for the very last.

I was very pleased when I heard of Dave and Teresa’s plans. Their land grab cut through my childhood stomping grounds and included my uncle's old farm.

Over the years, I have received numerous death threats. As leader of the movement to protect the planet's biodiversity, I am often blamed for lost profits by many of the world's power brokers—the Colombian drug cartel being the most dangerous. For safety reasons, I always travel under an alias nowadays and never fly commercial. Bob has seen to it that I have a long-range business jet with pilots at my disposal at all times.

I flew into Indianapolis International Airport on a Sunday morning, grabbed my rental car, and started driving. I was to meet Teresa and Dave at the property that once belonged to my Uncle Clyde and Aunt Julia. I had no trouble finding it, crossing the bridge over Lick Creek, and turning right up the gravel road that disappeared into the woods. I saw Teresa and Dave's van as I rounded the last turn. I parked beside it and got out. It was late fall and there was an icy chill in the wind. There was no sign of Teresa or Dave, so I decided to poke around until someone showed up. I hadn't been to this place since my childhood. The old barn was still there. The ponds had turned into meadows.

As I walked over the crest of a hill, I saw movement below. I pulled out my binoculars. It was Teresa and Dave all right, but there was a third person as well, holding a shotgun. I ducked down behind a tree and dialed 911 on my cell phone. The dispatcher put me through to the sheriff on duty in that county and I gave him detailed instructions as to how to find us. Fortunately, he wasn't far away. I continued to watch from my hiding place. It looked like they were being robbed.

The sheriff finally announced his arrival on my cell phone. I looked behind me and saw his partner and him getting out of their patrol car. I waved them over.

Teresa was sitting in her backpack on Dave's back. I remembered that she kept a voice controlled cell phone nearby at all times, usually with an earphone. It was worth a try. I called her number.

"Teresa here," she whispered.

"Teresa, it's Sarann. I'm watching you through my binoculars. Are you in danger?"

She nodded in the affirmative.

"I have two sheriff's officers with me. Tell Dave to set you down out of the way."

I watched as Dave set Teresa down. The assailant was clearly agitated by the unexpected move.

The sheriff fired his warning shot. The armed redneck turned and looked up the hill—biggest mistake of his life.

Dave is typically a quiet, shy man. I've noticed that a lot of really big men tend to be that way. Attached to a very slow burning fuse, however, is a murderous temper. Let the fuse burn long enough, and he will explode. He exploded.

By the time we got down the hill, the assailant was unconscious and bleeding profusely from his mouth and nose. I saw specks of bright white lying in the bloody leaf litter. I was aghast when I realized they were his teeth, roots and all. The arrest was made without resistance.

A few hours later, we were alone on the hill again. I said to Teresa, "You just had a narrow escape from the most dangerous predator on Earth."

"I see what you mean. My God, is this a common occurrence in Indiana?"

"Actually, anytime you hike outside of national preserves, you run the risk of stumbling into a meth lab.[64] Remember the movie Deliverance? I have had more than one uncomfortable situation in these hinterlands. It's over. Try to put it out of your mind."

"I'll try," she said, her tears starting to flow again. "You two can talk, I'm going to rest in the car."

Dave returned after helping Teresa into the van. "I am so sorry this all happened," I said. "Do you still feel like showing me your plans?"

"That's what we came for," he said as he rolled out a topographical map on the hood. "We plan to buy land to act as corridors to connect these state parks. We're getting this fallow farmland following the White River, the Wabash River, and these large creeks. A mark-up shows which roads we will kill. With careful planning, we can maximize our funds to get the most efficient ecosystem preservation possible. Eventually, we will have connection corridors between Indiana, Kentucky, and Tennessee. Here, take a look at this ad."

He pulled out a paper clipping that read, "Join the Wildlands Protection Patrol. Help us monitor protected Midwest ecosystems."

He continued, "This same organization is in place in Mississippi. We have private pilots who volunteer to fly reconnaissance missions and others who participate in two or three day patrols of the preserve boundaries. They're having fun and helping out. Best free source of manpower I've ever seen."

It was unusual to have Dave doing all the talking instead of Teresa.

"You OK, back there Teresa?" I asked, leaning toward the back of the car.

"I think so, just a little shaken up, being pregnant and all."

"Pregnant? Am I the last to find out?"

"No, actually, you're the first to know. We didn't want to say anything until it was a sure thing."

"Well, congratulations. I am delighted." I said as I reached to shake Dave's hand.

---

[64] http://espn.go.com/outdoors/hunting/news/2001/1204/1289816.html

He was grinning like the Cheshire cat. "We are going to name the baby after you if that is OK."

"I would be honored."

"By the way, your TIFIC worked perfectly," Teresa continued."

Looking at his watch, Dave interrupted. "We need to get back to Indianapolis. Before we leave we would like to invite you on a two-day patrol with the charter members of the first Indiana Wild-lands protection patrol. Interested?"

"Yeah, sure. I was planning to hang around here for a few days anyway."

"OK, here's a number to call," he said as he handed me a scrap of paper.

I spent that night in Indianapolis. I called the number Teresa had given me. The leader of the expedition was a woman named Marcy. She was from Mississippi and had lots of experience with these patrols.

I can see why these groups hold such appeal. The members view themselves as good guys fighting bad guys. This is the typical mind-set in almost all conflicts. In my opinion, there are no bad guys. Life is just a big power struggle. Timber, farming, and housing development can hardly be described as evil. We all live in structures made of timber and eat affordable food made possible by incorporated farms. There are just too many of us. That's all there is to it. Protecting what's left is just something that has to be done until our numbers shrink.

The next morning, I was standing at the hotel lobby waiting for my ride. Sticking to the advice Bob had given me earlier, I was using my youngest brother's name (Christopher) as an alias. I heard a horn beep behind me. When I turned, I saw a rusted out Ford van with faded lettering on the side that read "Mike's Meats."

I walked over.

"Chris?" The woman driver called out.

"That's me," I replied.

"Where's your stuff?" she asked with a raised eyebrow.

Bob had given me a Sentinel suit. I was wearing it with the helmet and power unit hidden in my backpack. The horns at the heels and knees had been removed.

"This is all I need," I replied.

"Didn't Dave tell you that we'll be hiking and camping for two days? You're gonna freeze your ass off in those… pajamas. Never mind, you can borrow stuff. Come on in."

I was to make sure the suit didn't fall into the wrong hands. Right—everybody in the world had, of course, heard about the suits. There were only a few thousand in existence, every one of them worn by an EPI ranger. They had been the subject of a number of news articles. Marcy didn't recognize it without the horns and helmet.

I clambered into the back. Camping gear and backpacks filled every available nook and cranny. Marcy had no idea who I really was or how I fit in with the EPI organization. I was doing this incognito. As far as Marcy was concerned, I was a friend of a friend

"This here is Chris, a friend of Teresa's and Dave's," Marcy told the group. "Chris is a little short on gear. I hope you all can find room in your hearts to share."

I met Buster, Mack, Jennifer, and Jackie. They were dressed in an assortment of spring, fall, and winter camouflage hunting gear and jeans. The green jungle camouflage on my suit did not seem particularly out of place.

We drove south near the Indiana-Kentucky border. Marcy pulled off the main thoroughfare and drove up a winding dirt road. We finally parked near an abandoned farmhouse.

"Everybody out. This is our starting point," Marcy bellowed.

I jumped out and started hauling gear with everybody else. Once everything was distributed into backpacks, we were ready to start. Two members of the team carried rifles. Two others were in charge of digital camcorders. We also had cell phones and a global positioning transceiver. One guy even had a pair of night-vision binoculars. The cold and damp weather was typical for late November.

My suit was running on its lithium-ion backup battery. It wasn't safe to use hydrogen as fuel in this kind of situation. The hydrogen gas that evaporates from the storage tank can be dangerous in a closed environment like a van. The technology behind the Sentinel suit was still an industrial secret. The U.S. military was developing a prototype, but the final product was still years away. Without my helmet in place, the suit's heating and cooling capacity was reduced, but the he battery still had enough power for this short hike.

As we trudged along, I noticed some sideways glances. I knew what they were thinking. "How could this idiot forget to bring camping gear?" The cold air was having its effect on my companions while I appeared totally unstressed.

The crew kept busy filming wrecked cars and documenting the global position of each. They also digitally recorded other things of interest, such as illegal trash dumps and out-houses.

This first reconnaissance mission was to be a survey of the newly acquired lands. By evening, everyone was pretty worn out.

My helmet was coming out tonight for sure. My ears were stinging from the cold.

The campfires were lit and everyone settled down to make dinner. I filtered some water from a nearby mud puddle, pulled a packet out of my pocket, and made a steaming hot meal of mashed potatoes with brown gravy, corn and peas.

"How did you do that?" Hank asked as I sauntered over to the fire shoveling food into my mouth.

I showed him my filter system and how the water could be made to come out of the filter boiling hot.

"That's one hell of a mess kit," he said with awe in his voice.

Everyone had gathered around. I figured this was as good a time as any, so I pulled my helmet out and put it on.

"God damn, that's a Sentinel suit." Hank bellowed. "I thought that only EPI rangers have access to those things. How did you get your hands on one?"

"I've got my sources," I said, smiling smugly.

"Is it true you can survive a direct lightning strike?" Hank asked.

"I seriously doubt it," I answered.

"Is it really bullet proof?" came the next in a long series of questions.

"No, it isn't. But you can wear lightweight body armor with it."

"Can you really sleep on the bottom of a lake?"

"No, but you can go under water for a short time in an emergency."

"Can you really read books and watch movies in your helmet?"

"Yep."

"Holy shit." Buster said with reverence. "Where does this suit of yours get all of its power?"

I showed him. "This is just the emergency battery. The fuel cell is the heart of a Sentinel suit. There's enough power in a half gallon of liquid hydrogen to run a fuel cell for a couple of weeks. I don't need it for this short trip."

"That just don't seem possible," Buster insisted.

"Think about it, Buster. Picture strapping a half-gallon carton of milk on your back. Liquid hydrogen has about three times as much energy as liquid propane. The average energy draw of a Sentinel suit is about equal to a furnace or stove pilot light."

"Hell, I see what you mean. I could probably grill for a month on a gallon of propane," Buster finally concluded.

After everyone had satisfied his or her curiosity, we settled down to a briefing from Marcy.

"One of our goals is to document abandoned cars and other eyesores for removal and cleanup. We also note the location of roads and trails for further analysis and possible removal. Wildlife sightings of special interest are also noted. These private property signs are to be put up at appropriate intervals, and we educate anyone we meet about the change of property ownership and land-use rules. Let's see... what else. Oh yes. We are to report any illegal or dangerous sightings like stills or methanol amphetamine labs to the sheriff. Once things get rolling, we will also deploy surveillance devices at strategic locations. Any questions?"

Just as she finished her sentence, the first snowflakes began to fall. The briefing was curtailed, and everyone rushed to get their tents set up as it began to snow harder.

I helped to set up the tents and tried to make myself useful in general.

"Chris, you're welcome to share our tent tonight," Marcy offered. "You're gonna freeze to death in that flimsy outfit."

"I'll make do," I said, trying to reassure her.

I pulled out my inflatable pillow, turned the heat up a few notches, and lay down under the stars to enjoy the storm while listening to some Beatles music. At first, the flakes would melt when they hit the helmet's faceplate but after a while it was completely covered. I fell into a dreamless sleep.

I awoke in the morning buried under a foot of snow. Someone was poking me with a stick. I sat up abruptly. Everybody had sticks in their hands. They had been prodding the snow as though looking for an avalanche victim. My emergency oxygen had started flowing automatically when the system detected high CO_2 levels. It would eventually have tripped an alarm if I used up my supply but, apparently, I had not been buried very long.

My sudden resurrection had shocked Marcy. "My God, Chris, I was sure you'd frozen to death."

Buster came to my rescue. "Marcy, I told you there wasn't nothing to worry about. Those suits are good to 'bout minus 100 degrees."

I kept myself busy getting fires going and preparing breakfast for everyone. Since I'd just spent the night in luxurious comfort I felt it was the least I could do. I even carried an extra backpack that day.

Over the course of the hike, we recorded the locations of thirty-seven junk cars and hundreds of dumpsites.

The signs we put up along the boundaries weren't the usual "No Trespassing" variety. They had been carefully designed and tested to gain the sympathies of visitors: "This is private land. It has been set aside for the privacy of wildlife. Please do not enter. Surveillance is accomplished with solar powered security cameras and motion sensors coupled to cell phone transmitters. You have ten minute to vacate this property."

We returned to our van at the end of our two-day hike, everyone but me looking worse for the wear.

On the ride back, Buster offered me $200 for my $60,000 Sentinel suit. I declined with the reassurance that his offer will still be there if I ever change my mind.

When the van reached my hotel in downtown Indy, I hopped out. Marcy got out with me and shook my hand.

"It was a great pleasure meeting you... *Sarann*," she said with an exaggerated wink.

It had been a fun adventure for all. Truth be told, though, the prevention of trespassing, dumping, and poaching on the preserved chunks of ecosystem cannot be continued forever. Our numbers have to start shrinking at some point. These preserved lands will all be surrounded by neighborhoods, roads, and strip malls soon.

Before leaving Indiana, I dropped in to wish Dave and Teresa luck with their upcoming adventures as parents. They were living in a large Airstream camper. The entire crew working on this new land acquisition project lived and traveled in a caravan of self-contained Airstream RVs. The idea was to not leave a footprint on the planet when they were done securing the preserve. No phone lines, power lines, sewer systems, roads or structures of any kind were built to support this contingent of real estate lawyers and conservationists.

"Enjoy your camping trip, Sarann?" Teresa greeted me as I entered. She had been working on the computer. Dave was busy preparing lunch.

"It was great. My Sentinel suit was a big hit."

"Where did you get that?" Dave asked.

"Bob gave me one," I said matter-of-factly.

"Did they know who you were?" Teresa asked.

"Marcy figured it out in the end." I said.

Dave chuckled, "The savior of the world's biodiversity crammed into a rusted out Ford van with a bunch of tree huggers."

The remark shocked me—savior of the world.

CHAPTER 13

The Siberian contract

EPI had secured a contract with a Russian conservation organization, and I was flying out for a visit. Few places on Earth have been more despoiled with industrial waste than Russia. Almost a quarter of a million people were forced to relocate as a result of the Chernobyl nuclear accident.[65] Chernobyl wasn't the first nuclear disaster, it was just the latest and biggest in a series. Iron ore smelters in central Siberia release two million tons of sulfur into the atmosphere annually. They constitute by far the worst single source of air pollution on the planet.

Siberia is the last place on Earth one should visit in the dead of winter. A more brutal history you will never find. Stalin's purges sent twenty million souls to their deaths here. Some say that there really isn't a good time to visit Siberia. The biting flies and mosquitoes make it a living hell in the spring and summer. The locals call them "gnusy" which translates into "the vile ones."[66] There is a brief window in the fall, before the brutal winter sets in, just after the first frosts subdue the biting insects, when a visit is feasible. My trip was strictly business. I would be depending on my Sentinel suit to protect me.

Russia is also one of the few countries in the world with a decreasing population. The causes are multiple. The average life expectancy for a male has dropped to 58 years, partly due to alcohol abuse. An increase in infant mortality, caused in part by pollutants in the environment, is yet another factor. Unofficial sources claim that about 60 percent of all pregnancies in Russia end in abortion. An impoverished woman is twice as impover-

---

[65] http://www.uic.com.au/nip22.htm

[66] http://forests.org/archive/europe/sibland.htm

ished when she gives birth to a child she can't provide for. Contraceptives are not nearly as available or affordable here as they are in other parts of Europe. Although abortions are risky, expensive, and painful, given the choice, women will use them to avoid unwanted pregnancies. On the other hand, the Russian people were glad to see the TIFIC arrive. The government provides both the contraceptive and the antidote upon request.

Unfortunately, AIDS has established its presence here with a fury. The successful implementation of an AIDS vaccine along with the TIFIC would be the greatest history-altering combination since the development of antibiotics and anti-viral vaccinations.

The news media are filled with stories of the Russian mob and institutionalized corruption. It's all very real and that is why I chose to carry my luggage rather than risk checking it. But, my own experiences with citizens of the former USSR have been positive. My Russian born car mechanic is highly intelligent, honest, and very hard working. The Russian immigrant who owns the small computer store in my neighborhood shares the same qualities. I've dealt with groups of Russian environmentalists and with some Russian basketball players, and I have yet to encounter a bad apple in the bunch. In general, they have gallows humor down to an art form.

My flights had been uneventful, and for this I was grateful. Russian aircraft, although superbly designed, are shabbily built and poorly maintained. I had been forced to fly my last leg in an ancient MI-8 helicopter. I had been warned to avoid doing that at all costs, yet, for lack of any other option, here I was sitting in the belly of one. In the aerospace industry, helicopters are sometimes sarcastically referred to as flying fatigue test beds. That's because they are subjected to high levels of vibration. Many aluminum parts will crack if given enough time. The trick is to inspect for these cracks and to replace parts before they fail. It isn't a question of whether cracks will develop; it is a question of when and where.

I was looking out of a round porthole at the frozen wilderness passing beneath me. The noise and turbine exhaust fumes had combined to give me a splitting headache.

Finally, our destination loomed up through my tiny aperture to the world, Vanavera, an isolated settlement of about 4500 in North-Central Siberia. This was as close as I would get to the Siberian tiger preserve using commercial flights. There are worse destinations in Siberia. Oimyakon, a town in Yakutia, often sees winter temperatures that drop below -90 degrees F.[67]

Conversation with my fellow passengers had been out of the question. We had landed and as the turbine engines wound down, I looked around me. I would not have thought this possible, but my traveling companions looked worse than I felt. One by one, we fell out of the helicopter into the biting arctic cold. As I passed by the cockpit, I saw what looked very much like a half empty bottle of Smirnoff.

I followed the person in front of me. We entered a makeshift terminal.

On the far side, I could make out a crowd of people convened to receive us. Everyone was wearing the standard cold weather garb: ankle length overcoat, thick mittens, and furry Russian hat. Everyone, that is, except Bob. Like a mime at a longshoremen's convention, the crowd was giving him plenty of elbow room. His baggy, snow white Sentinel suit looked absurdly flimsy.

"Bob!" I called out, greatly relieved to see a familiar face.

He walked over and gave me a heartfelt hug. "Lord to God, Sarann, I do not envy you your ride in that whirling death trap. I'm sorry to say this, but you're not quite done yet."

He threw me a backpack. "Here, put this on and meet me outside."

I looked around for a place to change. As I walked toward what I hoped was a restroom, I checked out the contents of the backpack. As expected, it contained a Sentinel suit. It would run for a short time on its emergency battery until I could hook up to some hydrogen. I put it on and went looking for Bob.

I found him outside standing next to an EPI ultralight that was fitted with skis. Steam was boiling out of its liquid hydrogen fuel tanks. Although the steam gave it the appearance of being hot, I knew that this hydrogen steam was cold enough to freeze-

---

[67] http://www.pcusa.org/pcusa/wmd/ep/resources/europe/rus1000.htm

dry my hand. Bob had donned his headgear and for good reasons. In the minute it had taken me to walk over to the ultralight, my ears were ready to fall off. I quickly put my helmet on also. Bob used hand signals to give me the correct radio frequency.

His voice crackled through my sound system. "Can you hear me?"

"Loud and clear," I called back.

It took a few minutes to connect an extra hydrogen tank to my suit.

"OK, let's go," he finally said, giving me the thumbs up.

I climbed into the back seat, fastened my restraining harness, and turned up the heat.

Bob pointed the craft's nose into the wind and took off directly from the taxiway. We climbed rapidly in the ice-cold air. While adjusting my noise attenuating system, I took in the view. There must have been at least three feet of freshly fallen snow.

I broke the silence. "So... what kind of safety record do these ultralights have?"

"As I've said before, there's nothing to worry about. The twin engines give us redundancy in case one engine fails. We wouldn't have enough power to climb with just one engine, but we can maintain level flight indefinitely which would give us plenty of time to find a safe landing site. They are also the only ultralights in the world that use fly-by-wire technology."

"What exactly is that?"

"Instead of using steel cables and pulleys, the control surfaces are moved by electric motors mounted directly on them. These, in turn, are controlled by a joy stick that provides input to three redundant computers."

"What if a wing falls off or something?" I asked, half joking.

"Look above you. That football shaped canister contains a parachute system. I can deploy it if we ever experience a structural failure.

"These ultralights are designed with the same triple redundancy you would find in a Boeing airliner. Jet aircraft use turbine engines, which are about a hundred times more reliable than piston engines. This is because turbine engines have just one major moving part—a shaft that has turbine blades connected to it in the back and compressor blades connected in the front. In addition, turbine engines have a simple rotational motion. Piston engines, on the other hand, have hundreds of moving parts that jerk back and forth violently thousands of times a minute. Our ultralights use rotary engines, which also have just one major rotating part. This makes them just as reliable as jet engines but much cheaper and quieter. Mazda used rotary engines in its cars for many years. In fact, these ultralights use scaled down versions of those Mazda engines. They have been modified to be lighter and use hydrogen instead of gasoline for fuel.

"These are also the only aircraft in the world that use fuel cells for their primary electrical system. Other aircraft use generators with conventional batteries for backup. The fuel cells tap into the same hydrogen tanks that feed the rotary engines. Again, because fuel cells have no moving parts, they are far more reliable than generators. Our suits by the way can also be connected to the aircraft's hydrogen tanks. For longer trips, it's a good idea to do so."

"OK, you've convinced me," I retorted. "Now, exactly, where are we headed?"

"We have a campsite about two hours from here. Ever hear of the Tunguska event?"

"Nope," I replied.

"In 1908, a comet fragment exploded in midair over this river with the force of a hydrogen bomb. Every single tree was flattened in a three-mile radius from the epicenter. People felt the blast's shock wave forty miles away. There was a glow in the sky that night which prevented darkness from falling. You can still see some of the old logs."

I was about to comment when, out of nowhere, I heard a voice singing, "*I'm a lumberjack and I'm okay…* " The hair on the back of my neck stood up. Adrenaline rushed into my bloodstream, and my heart began to race.

Bob's concerned voice appeared in my helmet. "What's going on? Your heart rate just tripped an alarm on the biorhythm monitor."

I was hearing voices again. The stresses of my travels were probably to blame. I began my calming, deep breathing exercises. My pulse rate quickly returned to normal. The voice faded into the background.

"I'm O.K… thought I heard something," I finally said.

Bob knew about the voices. "Monty Python again?"

"Yep."

"Which skit was it this time?"

"Never mind."

We flew on in silence. The Siberian forest stretched out to the horizon. My suit was keeping me quite warm although I suspected that it was at the extreme end of its operating capacity. The air temperature had to be at least 60 degrees below zero at our cruising altitude. The wind chill factor associated with an open cockpit ultralight was probably equivalent to 150 degrees below zero. Bob had explained to me once about how the suits were insulated. The mirrored liner was the main means of cutting radiation losses. Similar to what you would see if you peered into a glass thermos bottle, only flexible. Add to that the system of heat exchangers that extract all of the warmth from your breath before allowing it to exit, and you have a flimsy looking Sentinel suit that can keep you warm. The electrically heated fabric was the final touch, making up for where the insulation is inadequate.

We'd been flying for about an hour when the ultralight banked sharply to the left. "Can you see those snowmobile tracks?" Bob asked.

I looked down and saw the fresh trail winding its way through the forest.

Bob continued. "We're going to take a slight detour and see where those lead."

He changed course and began following the tracks. A few minutes later we came upon two snowmobiles pulling sleds. They were not aware of our presence due to the noise emanating from their own machines. Bob called in their global positioning coordinates to the base camp, and we returned to our original heading.

"We'll send a patrol out to determine if those guys are hunting tigers," he confided to me.

A short time later we began descending toward a small frozen lake. As the ground got closer I saw no sign of a camp.

We made a smooth landing and taxied over to the bank. Waiting for us when we arrived was a band of EPI rangers all perfectly camouflaged in their Sentinel suits. The camouflage pattern consisted of a white background with life size photographic images of tree branches and brush from the Siberian forest. An EPI ranger holding perfectly still forty yards away would be nearly invisible.

I was introduced to the group. Bob had imposed the same humility-instilling naming convention typical of his other ranger contracts. This time the group was using a sci-fi theme. I met Spock, Scotty, Sulu, and Chekov.

"So, what name do I get stuck with?" I asked Bob.

"This is a pretty large group. We have run out of the good names. You get to pick between Tribble and Uhura."

"Help me out here," I said. "You might find this hard to believe, but I have never watched a single episode of Star Trek."

"No problem. A tribble was a furry creature that multiplied faster than rabbits. Lieutenant Uhura was the communications officer on the bridge."

"I'll take Uhura then."

"I'll spread the word," Bob said, grinning broadly.

The tracks made by our ultralight were quickly brushed away. It was then pulled into the forest and hidden.

"Why all the secrecy?" I asked.

"We have reasons to believe that certain elements of the Russian mob plan to disrupt our operations."

"I see," I replied, realizing what I might have gotten myself into.

A convoy of camouflaged snowmobiles with trailers had been assembled. Bob jumped on one and I climbed on behind him. The first snowmobile pulled out, and we fell in line.

"How did you make these machines so quiet? " I asked as we moved through the woods in single file.

"What did you expect? The same way we quieted the ultralights—rotary engines with big-ass mufflers," he replied.

I should have known. Darkness had fallen by the time we stopped. Hours of daylight were limited this far north. I already knew from experience that there would not be a visible campsite. I was not disappointed. Bob walked over after helping to secure the machines.

"Won't you join Chekov and me for some dinner and conversation?"

"Delighted," I cordially replied.

We made seats out of big snowballs, placed them in a semicircle, and settled in.

I watched as Chekov stuffed snow into a heated pocket on his suit. At the same time, he pumped water out of this pocket with his electric-powered water filter, which also acted as an instant hot water heater. The boiling hot water was distributed into several heated containers of dehydrated food. I quickly followed his lead, and in just a few minutes we were having dinner in the hellish cold and pitch black night of Central Siberia.

"You guys must be going through hydrogen like there's no tomorrow," I said between mouthfuls.

"We are," Bob replied. "Everything is electrically heated. The food you just made would freeze solid before you could eat it if these containers weren't heated. A tank of liquid hydrogen, however, contains one hell of a lot of energy. We get by. These frigid temperatures force us to use a lot of hydrogen to keep things warm."

Chekov had finished eating and was working his way through a bottle of vodka.

"Why are you guys the only ones with these Sentinel suits? I asked. "Why aren't they commercially available?"

"Two reasons," Bob replied. "Cost and safety. These fuel cells and the liquid hydrogen storage systems they use are both very expensive items. In addition, using hydrogen is a risky thing to do. It is extremely explosive. We can afford this technology and are willing to take risks associated with hydrogen fuel. It's as simple as that. As a matter of fact, the suits have the potential to do a lot of damage to the environment should they ever become commercially available."

"Why?" I asked, astonished.

Chekov hadn't said a word yet. He was smiling, though.

Bob elaborated. "Just picture what would happen if every outdoorsman got one of these for Christmas. Ice fishermen, deer hunters, duck hunters, hikers. They would just suit up and walk off into the woods wreaking untold havoc. The hardships, discomfort, and danger resulting from extreme environmental exposure all act as a damper for would-be outdoor enthusiasts. Take those limiting factors away and you will have a brand new environmental disaster."

"I get the picture," I conceded. "You would add Sentinel suit exploration to the ever-growing list of legal outdoor recreational activities: trapping, hunting, fishing, hiking, camping, dirt biking, mountain biking, four wheeling, and brush picking."

"I get picture!" Chekov suddenly exclaimed.

I turned to face him. "You agree with Bob's premise then?'

"Da!" he replied enthusiastically.

I continued. "Have these suits made a big difference in your ability to patrol the preserve?"

"Absolute," he agreed.

He and I chatted in this manner for another fifteen minutes or so before he finally excused himself.

As he walked off into woods, I noticed Bob chuckling.

"What's so funny?" I asked.

"Sarann, Chekov doesn't speak a word of English."

"He doesn't?"

"Nope. He's always entertaining himself in creative ways like that. Come on, I have something to show you."

Feeling somewhat used, I followed Bob to a small white tent that I hadn't noticed before. It was well insulated and comfortable. Stretched out before us was a map of Siberia.

Bob started talking. "The situation here is more desperate than it appears at first glance. Russia has 25 percent of its original forests intact compared to 5 percent in the United States. People have the mistaken impression that Siberia is sparsely populated. In reality, there are almost 35 million people living here. Legal and illegal hunting and trapping go on year round.

"Ever pet a Siberian tiger?" he asked unexpectedly.

"No, can't say that I have," I replied.

"A male can weigh over half a ton. They are by far the largest cats on the planet—about two feet longer and a foot taller than a male lion. It is thrilling to be that close to an animal that might decide to kill you at any instant. Competition with hunters has created a relative scarcity of prey, but the main concern for Siberian tigers is poaching. Asian apothecaries pay a high price for any part of a tiger. A tiger's carcass is worth tens of thousands of dollars on the black market. Picture what must run through the mind of an impoverished hunter when a tiger walks into his gun sights while both are hunting the same deer."

"I see what you mean. How do you hope to control it?"

"Well, as you know, we were asked to help out here by the Russian Environment Ministry. A Russian environmental group called Zov Taigi—that translates into "Roar of the Taiga"—already has a good handle on the situation. Taiga is the Russian name for the continuous band of coniferous forests that cross Eurasia and North America. Siberia has many professional wildlife rangers who found themselves unemployed when the Soviet block disintegrated. That was when poaching hit its peak. Over seventy Siberian tigers were killed in 1993 alone. An American by the name of Steven Galster obtained funding to put the rangers back in action. They are called the Amba Patrol.[68]

"Fifteen men were selected from thousands of applicants. Minimum criteria included military field experience and the ability to maintain vehicles and communications equipment. These men were then given additional training and equipped with uniforms, Jeeps, weapons, and full field gear. Thanks to their efforts the Siberian tiger population is now holding at about 350 to 450. The problem is that these fifteen men are trying to patrol an area the size of Florida—about 50,000 square miles. That's why we were called in."

"These Amba anti-poaching patrols sound a lot like your EPI rangers," I suggested.

"Where do you think I got the idea?" Bob retorted. "They are going to need some help if we're going to save these animals from extinction. As a matter of fact, this is a training camp for the Amba rangers."

"Chekov is an Amba ranger?" I asked, surprised.

"You got it," he said.

---

[68] Note, most of the information in this paragraph and several others came from TIME Magazine September 4, 1995 Volume 146, No. 10 The Tortured Land by Eugene Linden/Yakutsk http://forests.org/archive/europe/sibland.htm

It was growing late. I found myself yawning at regular intervals.

Bob took it as a hint. "Time to turn in. Try to pick a place where a snowmobile won't run over you. I've assigned Spock here as your bodyguard. Try not to stray too far from him.

"Good night," he said as he walked off into the darkness.

"Well, Spock, how did you land that name?" I asked as I trod a flat spot in the snow for my bed.

In a thick Russian accent, he replied. "They tell me it was only logical choice. Don't ask me to explain it. What name was chosen for you?"

"Uhura," I said.

"The black woman in mini-skirt?" he asked, sounding surprised.

"What? I said, realizing that Bob had been entertaining himself at my expense as usual. "Never mind. Any advice on how to sleep in the snow in minus 50 degree weather?"

"Like Bob said, stay off beaten path. If it blows hard tonight, you will roll downhill. You can avoid this by lying with your head into the wind."

Spock picked a spot at the base of an adjacent tree and just lay down.

I inflated my pillow and settled in. I set my suit's temperature to 68 degrees and programmed it to cycle up to 80 degrees every 30 minutes. This gave me the cozy feeling of being wrapped in a hot electric blanket twice an hour.

The wind was picking up. It looked like we were in for a storm. Thanks to a geo-synchronous satellite, I had no trouble finding a good talk show.

I must have dozed off. I awoke late into the night to the sound of radio static but when I checked, the radio was off. That's when I realized the sound was being made by ice crystals blasting against my helmet's faceplate. The wind was blowing at least 40 miles per hour and gusting to 60 or 70.

I sat up and looked around. I could make out Spock's form off to my right. A storm like this was a life threatening experience for polar explorers of the past. For those of us warmly ensconced in Sentinel suits, it was interesting. Everything looked OK so I went back to sleep.

The next thing I heard was Bob's voice. "Rise and shine, everybody. We have guests coming for breakfast."

The sun was just rising. Bob continued with urgency. "Let's go, everybody, suits in cloaking mode. Let's go, let's go!"

I recalled what Bob had told me in Colombia. In this mode, the outermost layer of the suit is brought to the same temperature as the ambient air. Anyone watching through a heat sensitive scope would witness a vanishing act. The problem was that I did not know how to put my suit into this mode.

With panic in my voice, I said to Bob, "I don't know how to program for that mode!"

He came over. "Relax. I already took care of it for you."

"You did? I didn't know you could control my suit."

"There are a lot of things you don't know."

"So, what's going on?" I asked.

"Somebody in a surplus military helicopter is heading our way. I have to assume they aren't friendly."

As he spoke, I could hear the sound of an approaching helicopter.

"OK, Sarann, just hold perfectly still."

The helicopter arrived and circled for several minutes. It had a heavy machine gun mounted on the side with its operator clearly visible. Finally, it flew off toward the frozen lake where our ultralights were based.

"So far so good," Bob quietly said.

There was no need to talk quietly, but the fact that he did so strongly indicated the seriousness of our situation.

I activated the small video screen in my helmet. There was a surveillance camera watching the lake as part of our security perimeter. Twelve heavily armed men leapt out of the helicopter. They looked professional and very dangerous. Their first act was to destroy our ultralights.

"Look like ex-military," somebody said.

"I'd say ex-commandos," someone else volunteered.

Bob’s voice appeared again. "OK, here's the game plan. We can't use the snowmobiles to outrun them. The chopper will be able to detect our heat signatures with a heat-seeking scope. We can’t use the ultralights and with that helicopter around we wouldn't want to anyway. I want everybody to put on snowshoes. We are going to split up."

Bob, Spock, and I stuck together. Everyone else set off alone in different directions with instructions to hold at given global position coordinates. We started walking in a direction away from the approaching mercenaries. A few hours later we stopped to rest. Watching our video monitors we could see that they had arrived at our old campsite. They immediately destroyed our snowmobiles. I now saw why Bob had split the group up. The mercenaries had to decide which set of tracks to follow. They decided to follow ours. This is probably what Bob had intended. We were outnumbered and heavily outgunned, but they were blind. An EPI ranger is never alone. We can converse with each other and even transmit video images at will. We also know the precise global position of each team member. We began moving again as soon as the Mafia mercenaries resumed their pursuit of us. We could hike along in our warm Sentinel suits while they had to plod through the snow wearing their field gear and ammunition belts on their backs, losing energy to the cold all the while.

We set up new hidden security cameras along our route so that we could monitor their progress. Two EPI rangers had circled around behind the mercenaries and were collecting the hidden security cameras for later use. One of them, a marathon runner in civilian life, would occasionally arrive with a fresh supply of security cameras to deploy. We recharged the camera's batteries from our fuel cells.

We were also able to keep an eye on our attackers with our digital heat sensing scopes. When the outside air temperature is this cold, the human body appears as a glaring light in one of these scopes. At night, we could see their glow a mile away.

At one point, a mercenary finally spotted a hidden camera. We had to detonate it in his face. The helicopter made several attempts to drop soldiers in front of us, but we just changed course and skirted them. This cat and mouse game continued for two days and two nights. It finally dawned on our pursuers that they would never be able to catch us. While we remained fresh as daisies, they now had a casualty—from the exploding camera—and were suffering terribly from the cold. Finally, they gave up and were forced to climb a rope ladder back into the helicopter.

We rested and waited for the rest of our group to rendezvous.

"Why didn't we just shoot the helicopter down?" I asked Bob as we waited in the snow.

"Easier said than done. They were careful to keep it out of range most of the time. In addition, the engines are protected with thick armor plate and the blades are damage tolerant. We are on a training mission and do not have enough ammunition to engage in a protracted firefight. If we failed to shoot it down we would have given our positions away. They would have climbed to an altitude just out of our range and slaughtered us with their chain gun. Our munitions would be fighting gravity while theirs would be assisted by it. Body armor won't even slow down a round from a weapon like that."

"Oh."

Once we were all together, Bob briefed us. "The only option left is to walk back to Vanavera. Our antagonists have probably assumed that we will perish in the Siberian wilderness. It is highly unlikely that they're aware of the capacity of these Sentinel suits. That information is still a fairly well guarded secret. By my count, we have enough food and hydrogen left to hold up quite comfortably for another two weeks. It should only take us four days to get back to town. I expect it will be a very pleasant walk."

Bob was right. We emerged from the forest right on the edge of the Vanavera airport four days later. It was already dark. Sitting on the tarmac was the very helicopter that had attacked us six days earlier. Chekov said something to Bob in Russian and bounded off.

"What is Chekov up to?" I asked.

"He's going to make a minor mechanical adjustment to the helicopter's tail rotor."

"What kind of adjustment?"

"I think he's going to remove the Jesus nut."

"Jesus nut?"

"That's the one that holds the tail rotor on."

"Won't that kill someone?"

"Unlikely. They'll know it's missing as soon as they lift off and start spinning like a top."

We entered the town under the cover of darkness. Chekov had taken over leadership of the group. This was his territory.

"Where are we going?" I asked Bob when I caught up to him.

"We are heading to a bar where Chekov suspects he'll find the thugs that were chasing us around for two days."

"By the way, why were they chasing us?" I asked again.

"The Russian Mafia are behind the illegal trafficking of wildlife around here. That wasn't the first time they've tried to take out the Amba Patrol."

Bob, myself, and fifteen really pissed-off Amba rangers, arrived at the back alley door of a sleazy bar. The snow was filthy. A dim red light illuminated the scene. Chekov went in. We waited. A few minutes later he stuck his head out and gave the thumbs up sign. The rangers dropped their hydrogen tanks in unison and piled in through the door.

I was about to follow when Bob grabbed me by the arm.

Smiling, he said, "You do not want to go in there. You've never seen a bar room fight 'till you see how the Russians do it. The Amba rangers will be invincible with their Sentinel suits on. Body armor stops knifes and broken bottles better than it stops bullets. Headlocks aren't particularly effective either."

We left to find hotel accommodations. I was due to fly out the next day.

CHAPTER 14

San Diego

I was back in Seattle and sitting in front of the admissions desk for the Harborview mental-health facility. They were having a busy day and the room was packed. As I sat there, watching the mayhem, I had a flashback of a conversation that I'd had while at a dinner party years ago. The subject of mental depression had come up. A guest not known for having many deep insights remarked, "Depressed people have such a negative view of everything." I asked this guest if he had ever been depressed. Not surprisingly, he hadn't. This explained his complete and utter ignorance as to what depression actually entails. I on the other hand was all too familiar with the concept.

My thoughts were interrupted by the cheerful voice of the receptionist. "There are several people in line ahead of you. It's going to be awhile. There's some reading material in the corner over there if you'd like to find something to occupy your time."

I thanked her, and walked over to the pile of dog-eared and out-of-date magazines. A few books had also been donated to the waiting room or possibly just left behind. One of them was relatively new and a best seller. It was a big hardcover book. Opening it, I found that the text font was relatively large. It was almost as though the publisher was trying to make the book look bigger than it really was. It could just as well have been printed as a relatively thin paperback. What was it that had turned it into a best seller? Intrigued, I began reading. I am an exceptionally fast reader.

I finished the book two hours later and put it down. I realized in that moment that I was not the one who should be voluntarily checking in to a mental facility; a book written by a pair of evangelists about the anti-Christ battling God for control of the planet was on the New York Times best seller list. God had literally taken all of the good people up into heaven—titanium hips, pacemakers, breast implants, and all—leaving the sinners behind. The world had been instantly depopulated. Although not mentioned, those people left behind, the other creatures on the planet, and mother Earth herself must have all breathed a collective sigh of relief. As far as I'm concerned, the sooner that happens, the better.

The receptionist called to me. "I am so sorry this took so long. The doctor is ready to see you now."

"I've changed my mind," I said as I walked out the front door into the sunshine.

I called my assistant. "Randy, have my jet readied."

"Where you off to?" he asked.

"Sunny San Diego," I replied.

A few hours later, I was climbing into my Bombardier Challenger 300.[69] Bob had selected the Challenger jet for my needs because of its intercontinental range. My pilots were always ready to go on a moment's notice. Although I'd been back from Siberia for over a month, the chill had never completely worn off. I had a hankering to go someplace that was warmer than Seattle. It would give me a chance to revisit some habitat in that area and possibly learn a thing or two about the latest conservation efforts there.

Later, while gazing out the window from 30,000 feet, I was reminded of a discussion I once had with a young man of the Mormon persuasion. He thought it was inconceivable that anyone could believe the Earth was overpopulated. Coming from Utah, he was accustomed to looking out at vast expanses of desert. If you count the North and South poles, uninhabitable mountain ranges, deserts, and the oceans, you would see that only a small fraction of the Earth's surface has been populated by people. I guess, somehow, my Mormon friend got the word "overpopulated" mixed up with "overcrowded." It's a common mistake. He now has five children with plans for more. That's OK, people who want children should have the freedom to pursue their dreams. The avoidance of *unplanned* pregnancies is all that is necessary to bring our numbers under control.

---

[69] http://www.aerospace-technology.com/projects/bombardier/index.html#bombardier3

I took a long nap shortly after takeoff. I didn't wake up until I felt the jet banking for final at Brown Field.

First stop, Sea World. I didn't have any trouble finding the place. The parking was certainly ample and only cost seven bucks. Admission was forty dollars.

Would it live up to its reputation? The first thing I checked out was the freshwater exhibit. It had the standard circular layout. The crowds were thick. I found my place in the herd and slowly shuffled along. I was not impressed. After I popped back out into the sunshine, I fought my way to the shark exhibit.

As I moved lockstep in the line, I could look to my right and see sharks mulling about in a series of big concrete ponds. Finally, the line wound down under the ponds to a moving sidewalk that passed through a glass tube allowing us to view the bottom sides of the same sharks we just saw the tops of. Again, I was very disappointed. My impression was that Sea World was an amusement park using sea life for its props, replacing the bearded lady and strong man with trained seals and dolphins. It was all pavement, plastic, and long lines, carefully designed to extract every dollar possible from every customer. At one concession stand I was asked if I wanted to purchase a souvenir straw to go with my Coke. I declined, preferring to drink my beverage sans straw.

Two hours later I was fighting my way to the suspiciously difficult-to-find exit through thousands of overweight pleasure seekers wandering aimlessly about this circus with a nautical theme, as they stuffed their faces with cotton candy and ice cream.

Oh, well. There was still time to see the world-famous San Diego Zoo. It was located just a few minutes away from Sea World via the interstate system.

Upon entering, I was confronted by the same overweight, constantly eating sea of humanity I just left at Sea World. At least the grounds were beautifully landscaped with shaded footpaths.

There were posters all around the zoo touting the polar bear exhibit. What I found when I got there wasn't the bright white ball of cuddly fur I'd seen in the brochures. What I found was a thin, yellowed and deranged polar bear pacing back and forth in the classic throes of madness that are often induced in big cats and bears as a result of captivity in an environment that fails to adequately stimulate their brains.

Lying in the spacious and beautifully designed enclosure were several large, brightly colored balls that were the remnants of failed attempts to stimulate the poor creature. As it paced, it would swing its head just-in-time to miss the wall as it turned to precisely repeat the movement over and over again all day long like a broken machine.

Even the small children could sense that something was wrong.

"Mommy, why does it keep doing that?" I heard one little boy asking.

I tried to find the California condor display, but there was none. Years ago, the condor program was a media darling. I hadn't heard about it for a long time and this made me wonder.[70]

About twenty years ago, the Los Angeles Zoo captured the last fourteen California condors left in the wild. This move was prompted by the unexplained disappearance of six wild condors that were being observed and one more that was found dying of lead poisoning. At that point the total number of California condors on planet Earth, including those hatched in zoos, was twenty-seven. About ten years ago, eight condors were released back into the wild. Within a few years, four had died from collisions with power lines, one had died from ingesting antifreeze, and one had narrowly escaped being shot by a couple of yahoos with guns. Release of captive-bred condors continued. Three years later a milestone had been reached when, for the first time, some released condors found and fed on a sheep carcass. Up until that time the field crews had always fed the condors. Over the next few years the attrition rate continued, two more dying from power line collisions, two more being shot, and five others suspected of dying of natural causes. Why they assumed natural causes, I have no idea.

As the project approaches the quarter-century mark, there are a couple of hundred condors living in zoos and in the wild, all

---

[70] http://www.zooregon.org/Condors/timeline.htm

being fed by humans. Nobody knows how many millions of dollars have been spent on this project. If it were canceled today, the wild condors would be gone within a few years, shot, poisoned, electrocuted, starved to death, and run over. California condors can no longer exist on this planet without the direct assistance of people to feed them and protect them from other people. There are just too many of us. We are everywhere. There simply isn’t room on this planet for condors anymore.

The bald eagle was placed on the endangered species list and has made a remarkable comeback. People often use the success of the bald eagle as an analogy for the condors. It's a poor analogy. The total number of bald eagles in the wild never dropped below tens of thousands. They breed much faster and are far more adaptable. The elimination of the pesticide DDT was the biggest reason for their rebound.

As far as zoos go, I suspect that the San Diego Zoo is one of the best. However, a zoo is just a zoo. I found the Newly deer display particularly disturbing. A sign in front of it stated that this is the rarest deer in the world. It described the deer's marginal habitat and how unstable governments and warfare have recently made it extinct in the wild. It went on to say that a survey of all the zoos in the world found about fifty of them still alive.

That scenario, of a temporarily destabilized government and warfare hastening an impending extinction, is one that has played itself out over and over again, like a sporadic wildfire all across the planet. It has happened countless times in the past, and will continue to do so at an ever-increasing rate in the future. During the Great Depression, many zoos were forced to close their doors, and some even shot their animals and gave the meat away to hungry families. I have news for those people who think that zoos and their marine life equivalents offer a long-term solution for extinction—they don't.

The remains of hundreds of rare Tibetan antelopes were recently discovered.[71] They had been slaughtered to make expensive coats that are popular among the rich. Ancient cave paintings in France depict a now-extinct form of wild cattle called Aurochs. A small herd of these animals still existed in a private

---

[71] http://www.earthisland.org/tpp/LA_Times_science.htm

preserve in Poland until poachers killed the last one in 1627 AD.[72] In time, one species after another will be snuffed out in this manner. The Day is fast approaching when we read about the last remaining wild gorilla having been killed by poachers somewhere in Africa. I don't want to have to apologize to my children and my grandchildren if our generation allows this to continue.

I could not shake the feeling that there may not have been another human being in the crowd at this zoo who realized that the end of nature is truly upon us. In geologic time, the death throe will be over in just a few hours.

The infection that has afflicted Mother Earth may already be terminal. If the antibiotic I have given her takes hold in time, and if she has the strength to recover, then there is still hope in these last desperate hours.

I felt the icy grip of depression tugging at my mind again. I left uninspired. I planned to spend the night on the jet, pick up my brother in the morning, and check out some of the San Diego canyons before returning to Seattle.

The next morning I left early to retrieve my brother Pat. I hadn't slept well. My dreams had been plagued with visions of HIV infected children, extinct wildlife, and strip malls.

Pat moved to Los Angeles from Indiana when he was only seventeen years old. He stayed the first year with one of my sisters, caring for her child while she attended nursing school. After that, the family lost track of him. He would resurface now and again and let one of us know how he was doing. He finally settled down in a trailer in Topanga Canyon.

My brother is a very intelligent individual. He once helped me solve a problem that dealt with the mechanisms of failure in a chicken pox vaccine. We share a patent on the idea because of that conversation. I had always accepted his eccentricity as part of his personality but maybe there was more to it. He spent one winter living in an abandoned car, another under a highway overpass. He has explained to me the nuances and philosophy of begging for money.

---

[72] http://www.bigcats.org/esa/extinctions.html

I'd been driving for an hour by the time I arrived at his trailer. I noted that he had added some metal storage sheds to his property. I peeked into one and found that it was full of old computer components. As I approached the door to his trailer, I saw a coffee can full of cigarette butts. I knocked but no one answered. I tried the door and found that it was unlocked so I let myself in. There was fresh cigarette smoke in the air. Sitting in front of a computer terminal with his back to me was my brother.

"I'll be done a minute," he said without turning around.

I found a place to sit down. The room was cluttered, but not what you would call dirty. The smell of stale cigarette smoke permeated everything.

A few minutes later he stood up and gave me a wry smile. "Good to see you again. Have a nice trip?"

"I did, thank you very much. What are you working on?"

"Chaos," he said with that maniacal laugh of his.

I looked over at his computer monitor and saw the beautiful pattern of a Mandelbrot Set. I don't know much about chaos theory, but I know that a Mandelbrot Set is often used to graphically depict the idea.

"Interesting," I said. "Ready to go?"

"Yup," he replied. "Got any cash? I'm gonna have to fill up my gas tank."

"I think I have enough to cover us," I said as I realized we would be making the trip in his truck.

We threw our stuff into the back, and I climbed in the passenger side. It wasn't until we hit the highway that I realized there were no handles on the doors. There were armrests to pull the doors shut with, but there weren't even places for handles to be.

"So...Pat, where are the door and window handles?" I asked nonchalantly.

"I was wondering when you would notice that. I bought this truck from a guy who was a real control freak. Buttons under the dashboard on my side here control everything. They don't work, though, if you have a dead battery."

This led me to ask the question, "So what do you do if your battery goes dead?"

"I don't know," he said. "You sure can't kick the windows out because they're bulletproof. There's also half an inch of steel plating welded into the doors making them bulletproof as well."

We continued down the highway with me mulling over this information. My brother's truck was probably a remnant from the survivalist movement that swept through certain parts of the country in the '70s. It has always been my impression that Southern California contains more than its fair share of kooks.

We finally reached the outskirts of San Diego and pulled over next to a promising looking canyon. Although locally called canyons, they are actually big ravines, acting like natural drainage ditches. The zoning laws in San Diego prohibit building in these areas because of the potential for flash floods and mudslides during the infrequent rainstorms. Pat and I have always enjoyed bushwhacking and exploring together. We had hiked some of these canyons on previous trips to San Diego. I discovered as a result of those explorations that these canyons are, for the most part interconnected. This makes sense because they were created by centuries of water runoff. That is also why the city of San Diego has used culverts to keep them connected whenever a road bisects them. The canyons are an excellent example of how small ecosystems can remain intact as long as they remain interconnected with each other and with a larger ecosystem. The canyons do not enjoy any other kind of protective status. They were never conceived of as nature preserves or parks. They are a kind of no man's land frequented by adolescents sneaking off to smoke cigarettes and drug addicts seeking solitude.

It had been raining for almost two weeks before my visit. These rains were a welcome relief to the drought that had gripped this part of the country. The rains also gave me an opportunity to see some new southern California fauna.

As my brother and I descended into the canyon, we saw several large puddles. To our surprise, they were filled with tadpoles and crayfish. The peeping of frogs could be heard all around us. These canyons had been semi-desert on all of our previous forays, complete with cactus and sagebrush. Where had crayfish and frogs suddenly come from? As we hiked, my brother spotted an alligator lizard. There are alligator lizards in the Northwest but they are only about six inches long. This lizard was at least a foot long. My brother sneaked up behind it and the lizard immediately latched onto his finger. That had to hurt. My brother is an expert lizard catcher, but in this case, the lizard was an excellent brother catcher. The trick is not to look directly at the lizard as you approach it. You have to use your peripheral vision. It isn't magic. Lizards have evolved the ability to detect when a predator is looking at them. They are programmed to hold very still when it isn't—run like hell when it is.

Shortly afterward, I saw a roadrunner. We've all seen the cartoons of the coyote endlessly and hopelessly chasing the roadrunner—meep! meep! They are remarkable birds. They really do kick up a plume when they run down dry dusty roads. Because they are birds, they are very lightweight and can change direction with remarkable ease as they run. I've seen them make a ninety-degree turn without slowing down. This particular bird had a full head of steam and was heading straight for a tree. I cringed because it looked like he would run into it head first. Instead, he ran straight up it, just like in the cartoon. I'd never seen that before, neither had my brother. We looked at each other for a second and cracked up. So that's where they got the idea for the cartoon.

After exploring the canyons of San Diego, we were off to a state park. We arrived just as it was getting dark. My brother had made plans for us to camp out that night, but as it turned out, he hadn't brought any camping gear.

"This is southern California, you don't need camping gear," he reassured me.

So we spent that night lying on the bare ground without sleeping bags, pillows, or blankets. The stars were brilliant. I practically froze to death during the night but warmed up quickly in the morning. My brother can bear any form of physical discomfort as long as he has a cigarette in his hand. I had to envy him and his cigarettes for a while there. Breakfast consisted of a jar of peanut butter and an avocado.

We hiked for a few miles and then took a rest on a big boulder. My brother's endurance was limited because of his smoking habit. As we sat there watching a hummingbird buzz about our heads, a woman came jogging down the trail with not a stitch of clothing on. She came to a halt when she saw us sitting on the rock.

"Isn't it God's country?" she said with a look of rapture on her face.

"Yep, God's country," my brother calmly replied with a goofy grin on his face.

After she jogged on down the trail, he turned to me and said, "Welcome to California."

CHAPTER 15

The Congo contract

EPI has secured contracts from all around the world, Siberia, Africa, the Philippines, and Madagascar to name a few. The real effort to preserve the Earth's ecosystems started none too soon. Had the movement waited much longer there would have been nothing to protect.

EAI has been careful not to threaten oil supplies or other economic interests, and makes every effort to dodge the fist of the free market. Successfully protecting the rain forests of Colombia did wonders for the public image of EPI. However, there is a growing shortage of tropical lumber, and the problem is being handled very carefully. It was only a matter of time before the supply of lumber declined anyway. They have just started the process a little early while there was still some left. That's a rational argument. Unfortunately, rational thought has no more impact on instinctive status and profit seeking behavior than it has on religious extremism.

The Congo Basin is the second largest rainforest in the world.[73] These forests are disappearing at the rate of 15,000 square miles each year.

Some European lumber companies have decided that the price of tropical hardwood is high enough to justify the difficulty of logging in the Congo. They have begun a huge campaign of cutting logging roads into the jungle. These roads give hunters access to the forest. Combine these facts with the proliferation of high-powered rifles, and a wildlife slaughter has ensued. Even the workers at the logging camps supplement their meals with bushmeat.

---

[73]http://www.rainforestweb.org/Rainforest_Regions/Africa/?state=more

Bushmeat is the term used to describe wildlife that is shot or trapped for food or trade.[74] Many wealthy Africans prize the meat of Gorillas and Chimpanzees. Gorilla heads, hands, and feet can be seen in African markets routinely. The bushmeat trade has become commercialized with the meat being shipped all over Europe. Some estimates suggest that as much as 10 tons of African bushmeat may be exported to England every day. Some countries have poached almost all of their wildlife for export and there is nothing left. The extinction of wild Chimps and Gorillas is imminent. The wild chimpanzee population was in the millions just forty years ago. Their numbers have been reduced 80 percent.

Bob had asked me not to go to the Congo. He tried to warn me that conditions were presently more like a war than an environmental movement. Uncharacteristically, I chose not to take his advice.

As I stepped off the plane I was met by a full contingent of wary young EPI rangers. Although not wearing Sentinel suits, they all carried hidden weaponry. I was swept away to a safe location. We traveled only at night until reaching our destination in the Congo Basin. There I met Leto again.

"I'm sorry you chose to visit us at this time," Leto said to me.

"My apologies. I hope my visit doesn't become an unacceptable burden. I don't want to see my efforts degenerate into global guerilla warfare with no positive results for humanity. Your actions here have me a little concerned. Conflict may be a natural condition for humankind but that does not mean I will condone violence whenever I see it being used."

"I understand your concerns, Sarann. I hope your visit brings you some measure of reassurance. I am going to assign an entire platoon dedicated to your safety if you don't mind."

"That's OK with me. I won't be staying long."

On the ride out, Leto briefed me.

"Because of the Mogadishu debacle, the CIA has been providing us with clandestine military intelligence. The only catch was an assassination wish list of known terrorists. We are not, however, in the business of assassination and the list was dis-

---

[74] http://www.bushmeat.org/

creetly discarded. The CIA even offered the services of the elite Special Forces to assist with the terrorist hunt. Again, we politely declined. Although the Special Forces are superbly trained and dedicated, they still use conventional methods and weapons such as line-of-sight rifles, loud and easily targeted armored personnel carriers, daylight patrols, and helicopter support. Radio controlled mines are a dime a dozen. Armored personnel carriers have become sitting ducks. A child can place one of these mines in the middle of a road, run off into the hills with a remote-controlled detonator, and kill a dozen men with the push of a button. In addition, daylight aggression against an enemy that has limited night vision technology is dangerous and wasteful in the extreme. The foolishness of exposing one's head and upper torso just to aim a rifle is an idea our weapons have made obsolete.

"Their reliance on helicopters is also unwise. A helicopter rotor will be destroyed in most instances if it makes contact with a fist-sized rock. In other words, a helicopter can be knocked out of the sky with a rock, literally. Complicate that scenario with today's shoulder launched rockets and only a fool would commit helicopters to close front-line combat support. The exact location of an enemy force can be pinpointed just by watching where the helicopters land—the most valuable piece of intelligence an opposing force could hope for. We will have nothing to do with such vulnerabilities.

"The years of unending warfare have bred a vicious, psychopathic subculture of brutal bandits who are extremely difficult to deal with.[75] Well-armed and violent warlords rule this region. We have a weapons buy-back program. Anyone willing to sell his AK-47 will be financially sound for several years unless he blows it all on whisky.

"We have made some technological advances since the days of the Colombian contract. For example, we have enhanced our communications capabilities with communications balloons. Similar to weather balloons, they carry special relay hardware and are tethered to the ground with a strong line. They are essentially mobile, 1,000-ft. tall towers. Communication is a ranger's main strength. Everything else is secondary."

---

[75] http://www.cnn.com/2003/WORLD/africa/04/10/congo.facts.reut/

"Aren't the balloons pretty easy to spot?" I interjected.

"Only in the daytime. We reel most of them in at sunrise. We use some balloons like duck decoys. Anyone trying to neutralize them finds out they have stepped into a meat grinder for nothing. This allows us to mix some active ones in with the decoys during the day. Everybody has pretty much learned to leave those balloons alone."

The analogy that crossed my mind is a boxing match where one of the combatants has to wear a blindfold. "Weren't you using a combination of satellite relays and cell towers for communicating in Colombia?"

"That's right, but the cell towers were too vulnerable. They were too easy to locate and too difficult to defend. In addition, the U.S. military got nervous and has not allowed anyone to sell us satellite bandwidth for months. Staying one technological step ahead of your enemy is part of the game. Wait till you see the new Sentinel suits."

I could just picture what it must be like to fight an opponent that has no base of operations. A contingent requiring no tents or shelters, where each EPI ranger lives in a self-contained, armored, Sentinel suit and is in constant contact with each other. Technology had outstripped tactics during the American Civil war. The accuracy and range of new cannons and rifles caused tens of thousands of casualties before the military minds woke up and stopped marching their armies at each other in straight rows. With the advent of the Sentinel suit, technology has again outstripped tactics. The analogy to me would be a species of hornet that, instead of living together in a nest, all live solitary lives but can come together to defend their territory or attack an enemy and then disperse again. Trying to eliminate them would be very difficult indeed.

"I wish we could come up with more economic incentives to prevent the destruction of these rainforests," I said. "Force appears to be the only immediately available solution until we do. Hopefully, that won't be for long. My contraceptive distribution organization is in this area. They have a plan to provide a year's supply of canned meat to those who are willing to take the TIFIC. The idea is to eliminate the need for protein supplements to their vegetarian diet and hopefully decrease the pressure on bushmeat hunting."

Looking solemn, Leto said, "The idea won't affect the bushmeat problem. Most bushmeat is traded. The hunter is not usually the one to eat it. Most of your canned goods will just be traded as well. These people are looking to escape poverty, not to have a balanced diet. Yours is the long-term solution, Sarann. These Boy Scout exercises are only meant to be a temporary holding action as you well know."

Shortly after our arrival at Leto's base camp, I was handed my new Sentinel suit. The only thing that *looked* different was the helmet. The visor was camouflaged to match the rest of the suit instead of being mirrored. I put it on and found myself in the dark. The visor was completely opaque; I couldn't see a thing.

I opened it up so I could see. "Hey, what's the deal with the helmet? What good is a visor you can't see through?"

"We have made the visor bullet proof," Leto said. "You can take a round right in the face now and live to tell about it."

"It seems to me," I continued, "that not being able to see was a hell of a compromise to get a bulletproof faceplate."

"Your sense of humor agrees with me, Sarann. The visor now uses a system of compensating optics and magnification mirrors. Power your suit up before you close the visor."

I took his advice. He was right. Upon closing the visor the camera that was integral to the helmet became my eyes. Everything it saw was projected on the inside of the opaque faceplate. It automatically adjusted for light conditions, and I could zoom in on breadbox sized object a half-mile away. An opaque faceplate made the Sentinel suit even more sinister looking—not that it needed to be more sinister looking.

About the time I had finished familiarizing myself with the new suit, a series of sensors along the main dirt road detected a large paramilitary contingent heading our way. Sentries made detailed intelligence reports describing the size and lethality of the force. They were moving too rapidly to wait until nightfall. This action would have to be staged in broad daylight. Every ranger seemed to know what to do. The communications channels were abuzz with commands and instructions. The platoon assigned to protect me jogged to their assigned positions. Every available hand would be needed. As usual, I was sequestered out of harm's way with a bodyguard, but at least I had a good vantage point from which to view the unfolding ambush.

The advancing contingent had recoilless rifles and lightly armored personnel carriers among the trucks that carried most of the troops. One of my bodyguards informed me that the armored carriers would be targeted last. He also gave me a channel I could tune into to hear some of the battle language as it flowed back and forth between commander and ranger. I could not make heads or tails out of it. As usual, most of it sounded like gibberish.

I had been watching this scenario for about a half-hour when I heard a series of popping sounds. I saw dust spring up around the advancing caravan. Several trucks veered off the road and the sound of enemy gunfire erupted.

My bodyguard briefed me. "That loud pop you heard was the sound of dozens of rocket propelled weapons going off simultaneously in starburst patterns from long range. Most of the contingent's vehicles were destroyed in that second. Every tire, every engine block, and every gas tank looks like Swiss cheese at this point. Hopefully, there was not much collateral damage. From this far away, the bullets have time to accelerate to supersonic speeds. The sound of them hitting metal is followed about one second later by the sound of them screaming through the air. The gunfire you're hearing now is coming from the warlord's weapons. They are shooting at ghosts. We will not fire another shot until the time is right. Most EPI rangers are now redeploying into a new position along the road in the direction this militia just came from. By the time they collect their wits and run for home, a new and more deadly ambush will be waiting for them. Come, follow me."

He jumped up and began jogging with me right behind. We moved along the ridge paralleling the road below for about 10 minutes before we rested. With my bodyguard's assistance, we were able to spot about a dozen other EPI rangers lower down the hill from us. We also spotted several rangers planting a series of radio controlled mines in the road.

We waited. As expected, the panicked remnants of the convoy came racing back down the road. At this point, it consisted almost entirely of vehicles too heavily armored for the rocket bullets to immobilize. The mines were detonated at precisely the right times under the engine compartments. This ability to detonate a shaped charge into the engine block wrecks the vehicle while sparing the occupants. Every war machine had been totally annihilated in less than an hour. Those that could, were running back down the road as fast as their feet would carry them.

EPI rangers descended into the valley to do what they could for the injured and to collect dropped weapons.

I wanted no part in the aftermath of this conflict. I lost my bodyguard and headed back to camp unescorted. A short time later, as I rounded a bend in the trail a bullet struck me in the chest with the force of a tire iron, knocking me flat on my back. I felt like an NFL lineman had just tackled me. Lying in the mud, I looked up to see four khaki-clad Congolese soldiers bristling with grenade launchers and automatic weapons. I thought their eyes would pop out of their heads when I stood up unharmed. My body armor had stopped the bullet cold. It took a few seconds for our mutual surprise to wear off. They kept their weapons leveled at me. It was obvious from their reaction that I was the first person they had ever seen in a Sentinel suit. I knew they wouldn't be able to hear my voice from inside the helmet, and I quickly initiated a mayday call.

I heard Leto’s reassuring voice through my headset. "Try to remain calm. We’re on it."

My captors gestured for me to remove my helmet. The heat and humidity hit me like a baseball bat. Their relief was apparent when they saw that it was an ordinary person inside the suit. They secured my hands using baling wire. As we began to walk rapidly down the trail, I made a mental note to myself; this was the last time I would visit an EPI security zone.

We had been hiking for over an hour. Without the use of my hands, I couldn’t brush the insects away. The flies, ants, and mosquitoes had zeroed in on my exposed head, and sweat was blurring my vision. There was no sign yet from Leto. Things took a turn for the worse when we stepped out onto a logging road and got into a waiting truck. We sped away leaving my hope for immediate rescue behind.

I had grit in my teeth and road dust covering my face when we finally arrived at a village. The headman was sitting in a new Range Rover, looking at maps and talking over a radio. He looked up in time to see me being taken into a large tent.

Smiling, he entered the tent behind me. "We have succeeded in capturing a ranger, and not just any ranger. You are Sarann, are you not?"

Silence.

"This is too good to be true. Hostages and blowtorches make a powerful combination," he said as he left the tent.

I was in deep shit. Leto must have figured out by now that the offensive had been a ruse to allow a team to go in through the back door and take a prisoner. I carried no weapons. Capturing an armed ranger would have been a lot more problematic. I had been in a similar pickle once before. Fortunately, out of curiosity, I had taken some combat medications just minutes before my capture. Thanks to those medications, I was not nearly as concerned as I should have been. I sat alone in the dimly lit tent with the guards. By the time the headman returned, night had fallen. This had given me time to formulate a plan.

As he walked in, I calmly asked, "What do they call you?"

"They call me Dada, because my great uncle was Idi Amin Dada—butcher of Africa."

"What should I call you? Great nephew of the butcher of Africa, or Dada?"

"Dada."

"Well, Dada, you have gotten a lion by the tail. You will have to let go of it carefully if you are not going to get bitten."

At first my flippancy appeared to anger him, but I detected a change in his facial expression as I smugly stood my ground. I had successfully planted a seed of doubt that I hoped to capitalize on later.

"We will see who gets bitten," he countered. "What information would you like to volunteer before we begin to interrogate you?"

"I'll start by telling you that Leto is listening to our conversation with great interest."

The mention of Leto's name gave him a start.

"And just how is he doing this?"

"Through a radio implanted inside my body," I said.

He walked over to a blowtorch and lighted it. He was counting on the sight of a lit blowtorch to send me into a sweat popping panic, but I didn't flinch. These combat medications were not like anything I had ever experienced. I had no fear. I calmly mused over their pain killing capabilities. I thought it best to remain silent as he walked toward me.

"Can he talk to you as well?" he said, stopping short.

"Oh yes. He wants to know how much you value your life."

"You are lying to me," he countered. "I could gag you."

"You could, you imbecile, but how would I answer your questions?" I said with as much disgust as I could muster. This act of bravado seemed to give him pause. I had been told that the combat medications made one's skin completely numb to the pain of burning, although one could still sense pressure. Maintaining the ability to feel pressure was necessary in order to walk. I was taught in medical school that people who lose all feelings in their legs soon break their ankles and dislocate their knees because of the lack of sensory feedback.

Dada ordered a guard to hold out one of my exposed hands. He then passed the torch across my knuckles. He looked up into my face expecting to see me wincing with pain.

Instead, I smiled at him and in doing so, watched his smile disappear. Seeing the first glimmer of fear in his eyes, I pressed home my advantage.

"Leto is near," I said. "He and his rangers can see your heat signature through this tent like it is made of glass...."

I was interrupted by an internal phone call.

"Yes, I understand." I said to the air.

Silence.

"The one holding the torch." I said.

More silence.

"I will tell him." I finally said.

Dada, listening in on my half of the conversation with Leto, had dropped the blowtorch.

I continued, "Leto has a proposal for you."

"What is it?" Dada asked, looking worried.

"Have a guard escort me to the edge of the jungle and Leto will let you live."

Dada swallowed hard, thinking over his options. "How do I know he will not kill me anyway?" he finally said.

"You have my word," I replied.

He said something to a guard. I stepped outside the tent and walked with the guard in tow, past the other soldiers and their campfires to the edge of the jungle. The guard stopped and I continued to walk off into the dark. There was a tense moment as I wondered if the guard had been told to shoot me in the back. I soon found myself alone in the pitch-black Congo night. I had done it. I had completely bluffed my way out of the hands of a murderous warlord by pretending to have a radio hidden somewhere in my body. All those years of getting my butt waxed in poker games by the other think tank members had paid off. Where in the hell was Leto? How was he going to find me if my global position transmitter was with my confiscated helmet? I decided the best thing to do was to continue walking down the dirt road in the dark. If there were rangers about, they would find me soon enough.

I was extremely thirsty, but not thirsty enough to risk a drink from standing water. As I walked, my thoughts strayed. Guinea worm disease is contracted by drinking water contaminated with Dracunculus larvae. The larvae are swallowed by tiny water fleas. Once the worms have matured inside the water flea, any person who swallows water containing one of these minute fleas becomes infected.

For the next year the Guinea worm grows to a full size adult. They are as thick as spaghetti noodles and can grow up to 36 inches long.

The worm will then migrate to the surface of the body. A blister develops on the skin where the worm will emerge. When the blister ruptures, it causes a painful burning sensation. The victim will often immerse the blister in water to obtain relief. This causes the adult female to partially emerge from the wound and release a milky solution containing millions of immature worms into the water. The Guinea worm will continue to release worm larvae whenever it is exposed to water for the next several days.

In addition to the unending warfare, the Congo is also home to an unending list of diseases—the most terrifying being the

Ebola virus. Named after the Ebola River where it was first identified, this pathogen usually kills its victims a few days after symptoms first appear. 70 percent of the people who catch Ebola, die from it. It is still unknown as to where the virus comes from. Particularly gruesome are the victim's last days, in which they leak blood from every orifice. The victim also develops subcutaneous bleeding, creating a discoloration just under the skin, and turning them bright red. A person who has died from this horrible disease is a site to behold—bright red in color and lying in a pool of blood.

Other diseases include, Yellow fever, Dengue fever, Malaria, filariasis (elephantiasis), leishmaniasis, onchocerciasis, trypanosomiasis (African Sleeping Sickness), Cholera (black death), Schistosomiasis and Guinea worm disease.

A worm also causes filariasis, commonly called elephantiasis because it causes ones arms, legs or testicles to swell to gigantic proportions. This worm is so tiny it is transmitted via mosquitoes. There are few sights more unsettling than a man carrying his own testicles about in a wheelbarrow. It has no cure.

The combat meds were wearing off fast. The burn on my hand was growing painful. I was exhausted and starting to feel anxious like a normal person again.

I was startled by a voice in the dark, "It was dangerous of you to leave the protection of your bodyguard."

It was Leto.

I looked up to find that I was surrounded by rangers. They immediately began looking after my concerns.

"Let's get something on that burn."

"Take these pain pills."

"Drink this."

"Are you hungry?"

I turned to Leto. "This is not my cup of tea, I think I've seen enough."

"This fighting will not last much longer," Leto said. "Once the warlords learn to stay out of our weapons free zone, they will cut their losses and focus their attentions on each other again."

"You know," I said, "by definition, *you* are a warlord. Keep that thought in mind."

He looked at me without expression, turned, and started down the trail.

"Hey, wait a minute. I need a helmet." I called after him.

"We'll get you a new one," he said without turning around.

The next day Leto sought me out. "I tried to warn you. These are not good times. No amount of negotiating can bring order to this land. There are tens of thousands of bored, unemployed young men roaming this part of the world, armed to the teeth. The quickest way to bring order back is to forcibly disarm them and bring to justice any leaders who refuse to surrender their power base. It is a violent solution but a necessary one. Come, I want to show you something."

I followed him down a winding path to the edge of the main road where a large semi-tractor-trailer sat. He flung the doors open. The truck was filled floor to ceiling with thousands of deadly weapons that had been confiscated or purchased through the buy-back program.

Next to the trailer, half a dozen EPI rangers were watching a computer screen. As Leto and I walked up, I could make out a video camera image. The live image looked as though the camera was being handled by a group of men. None of them seemed to realize that the camera was turned on and broadcasting their images as they passed it around.

"What's this?" I asked.

"That image is from the camera in your captured helmet."

Leto nodded his head, and one of his commanders pressed a button. The screen went dead. A few seconds later I heard the muffled sound of a distant explosion.

"Must be closer than we thought," Leto mused out loud.

That's when I realized the distant rumble had come from the explosive charge in my captured helmet.

"By the way," I said, "I promised Dada that you would spare his life."

"You should have said something sooner," Leto replied.

As nightfall descended, EPI rangers began appearing out of nowhere, as they are wont to do. We were going to regroup, rest, and head out late tonight. We had two walking wounded. They would be flown out tonight. Under cover of darkness, a couple of ultralights with floats were being assembled along the riverbank.

It took us two days to hike to our re-supply area. It was time for me to go. I was escorted to a Cessna 206 parked on a grass airstrip. We flew to the nearest airport capable of handling my business jet. The jet had been circling and landed right behind us. Seconds after I disembarked from the small plane, the jet taxied over with the door open and stairs hanging out. It stopped just long enough for me to clamber on board and we were off.

CHAPTER 16

The pear

I was sitting in a traffic jam, in a filthy cab, in the smog-filled middle of Mexico City. I couldn't take it anymore. I paid the driver, stepped out, and started walking. Jesús had extended an invitation for me to visit him in his Church, which also served as an orphanage for street children.

The air was oven hot. A strong wind had picked up. Mexico City is one of the few places in the world where the air quality is so bad you can actually inhale a gastrointestinal disease like hepatitis or dysentery—normally diseases contracted in contaminated water.[76] Keeping a handkerchief over my face, I had been walking for about 30 minutes, keeping to whatever shade I could find. I rounded a corner and there it was, looming up out of the smog like an ancient stucco Castle. The doors were huge and made of solid oak. I grabbed one of the oversized doorknockers and let rip.

In the middle of my second attempt the door swung open. Half expecting to see a hunchback, I was relieved to find instead, an attractive, well dressed, young woman.

"Hello. I'm here to see Jesús," I said.

There was an uncomfortable moment while she stared at me with her mouth open.

"He should be expecting me." I reiterated.

Finally she spoke. "You are the one? You are Sarann?"

"That's me," I replied, not knowing what else to say.

She gestured for me to follow. The interior was even more castle-like than the exterior. We wound our way up some narrow, dimly lit stairs. They ended at a single door. The young

---

[76] http://healthandenergy.com/mexico_city_air_hurting_kids.htm

woman knocked. The door opened and a silhouette appeared. The bright light from the room contrasted starkly with the dark stairway.

"Sarann, good to see you again. Come in," Jesús declared. "Maria, could you please turn the air conditioning down a little? It's colder than hell out here. Turn the lights up a notch while you're at it. Thank you so much."

I walked into a cool, brightly-lit room lined with bookshelves.

"Welcome to my office. Can I get you something to drink while we blast dung?"

"I think you mean, 'shoot the shit,' and yes, anything would be fine to drink," I replied.

As he busied himself in the small kitchenette, I began my usual litany of questions. I like asking questions and listening to the answers. It's a propensity that has served me well through the years.

"Maria practically went into shock when she realized who I was. What's with that?"

"I've told her a few things, that's all," he said.

"What kinds of things?" I asked as he handed me a glass of juice.

"Just things, Sarann," he continued.

He pressed a button on the intercom. "Maria, two for dinner tonight."

"Yes, my lord."

I almost choked on my drink. "Yes, my lord?"

With a look of bored resignation, Jesús rolled his eyes and said, "Maria is convinced that I am the Second Coming."

"Are you?" I asked in jest.

"I wish," he said flatly and immediately changed the subject. "Maria is an absolutely awesome cook. You are in for a treat tonight. Come, let me show you my sanctuary."

"Lead the way," I said as I stood up.

We wound our way back down the stairs. Although it was dark, I could tell that we had entered a large chamber.

"Lights on!" Jesús suddenly called out.

Revealed before us was a huge amphitheater designed along the lines of traditional Catholic Churches. It was beautiful.

Behind the podium were several glass display cases containing various holy relics. Among them, the blue raincoat that had been the subject of a sermon I had listened to.

"How does an object qualify as a holy relic?" I asked as I squatted down to get a closer look at a pair of ancient wrist shackles.

"The main criteria is authenticity. The story behind the relic must be backed up with solid, verifiable evidence," Jesús said.

"What's the story behind these?" I asked, looking up.

"They once bound the hands of a victim of the Spanish Inquisition."

"Well?" I prompted.

He continued. "A graduate student of mine had stumbled upon some interesting information while doing research on the Inquisition. In 1600, the same year a Dominican friar by the name of Giordano Bruno was burned at the stake for insisting that the Earth circled the Sun[77], a woman by the name of Mercedes had been bound to a stake atop a pile of firewood. The fire had been set, and as was customary, she had been given one last chance to convert. She didn't do so until the fire was almost upon her. As was also customary, she was then strangled by the executioner before the fire reached her as a reward for conversion."[78]

I was appalled. "I had no idea the Church had such a brutal history."

"Oh, yes. A detailed ledger of each execution was kept, recording the names of the victims, whether or not they recanted, and any other details of interest."

"Is there more to the story?" I asked.

"Oh, yes. This is where things get interesting. According to the records, a rainstorm swept in that day extinguishing the fire that was meant to consume Mercedes. Some of the executions had to be rescheduled. Mercedes, having already been strangled by the executioner, was given a proper Christian burial thanks to her last minute conversion. Coincidentally, another student of mine was doing research on yet another woman by the name of

[77] See Galileo's Daughter by Dava Sobel, pager 4
[78] http://www.tyndale.org/TSJ/6/daniell.html

Mercedes who resided in Paris in about the same timeframe. She was the founder of an international orphanage for abandoned infants that exists to this day. I thought this an interesting coincidence and asked my students to try to find more information. Further research turned up portraits of both women. Although one painting had preceded the other by ten years, it was obvious that they depicted the same person. The portraits were nearly identical except in one, the woman was a decade older and had badly disfigured hands and forearms."

"Where did you get the shackles?"

"We made a field trip to Madrid. After a week of searching, we finally found Mercedes' tombstone. Her name and the date of her death were still barely legible. I successfully petitioned the State to exhume the grave. Aside from these shackles, the coffin was empty. Because of Mercedes' hesitancy to recant, and with the fire closing in, the executioner had done a hasty job. She had been asphyxiated only to the point of unconsciousness and inadvertently buried alive."

A chill spread down my spine. "How did she escape?"

"Look closely at the shackles. Part of the locking mechanism was made of lead, which has a very low melting point. The fire had melted those parts allowing the shackles to fall open. Ruined by the fire, they were probably thrown in the casket along with her body."

I could just make out the shape of melted metal extruded from the keyholes.

"That would explain the burned hands and forearms." I speculated.

Jesús continued. "There is no question that she escaped from her coffin. It may have been left unburied until the rains abated, giving her time to flee into the darkness. We'll never know. Mercedes is our first Saint by the way."

In the next display case, another object caught my attention. It was made of brass and roughly the size and shape of a stage microphone. The figure of Satan was cast at one end.

"What's this?" I asked, pointing to the strange object.

"That, Sarann, is a torture device called the pear.[79] It was used during the Inquisition to force heretics to confess that they were not of the Catholic faith."

"How does it work?" I made the mistake of asking.

Hesitating slightly, he continued. "It was forced into the victim's mouth, anus, or vagina by the Church-appointed torturer and expanded with a screw mechanism until the body cavity was ripped asunder. Come, dinner should be ready."

Food was the last thing on my mind at that moment.

As we entered the dining room, the aroma of Maria's cooking soon chased my nausea away. We sat down to a meal of pan-fried Chilean sea bass with chipotle butter, asparagus shoots, Mexican rice, and the best Chardonnay I've ever tasted.

Maria hovered in the kitchen, stealing glimpses as we dined.

The conversation soon fell into the standard pattern, me asking my host questions while I sat back and savored the responses for their intellectual content.

I started. "Isn't Chilean sea bass on the endangered species list?"[80]

"Are you serious?" Jesús said, flabbergasted.

"Actually, no. Just thought I'd jerk your chain. Some zealous environmentalist group inadvertently started an urban legend that they are on the endangered species list when they used the word endangered in a press release. This group was just trying to stop the collapse of yet another fishing industry primarily due to illegal harvesting. These fish aren't on the endangered list—at least not yet— and they also aren't related to bass. The name sea bass is purely a marketing scheme to make them sound more appealing. Their common name is toothfish. They can live up to half a century and can weigh two hundred pounds. Ironic choice for our meal wasn't it? We may as well enjoy, it being cooked and all that."

"I was unaware," he said. "I apologize."

"Stop worrying about it. The problem isn't so much what we eat, the problem is that there are too many of us eating."

---

[79] http://members.fortunecity.com/babygurl1385/#The%20Pear

[80] http://www.nmfs.noaa.gov/trade/chile.pdf

Changing the subject, I asked, "What exactly was an Inquisition?"

"Heretics were regarded as enemies by the Roman Church. The first in a series of Inquisitions began when Pope Gregory IX instituted the Papal Inquisition in 1231 for the apprehension and trial of heretics. Suspects were tortured to make them confess to heresy and eventually handed over to civil authorities for punishments like burning at the stake. The Church preferred burning because it did not shed blood. They had a thing about shedding blood. It is rather comical to search the Internet on the subject. Many authors try to distance the Church, claiming that the Church had very little if anything at all to do with the Inquisitions. This is ludicrous. Coincidentally, these authors are most always members of the Catholic religious hierarchy. In their estimates, only a few thousand people were burned at the stake—tops. It was no big deal. Everything was the government's fault. I'm sure the same attempts to revise history will be applied to the current sex scandals.

"On the other extreme, there are some fundamentalist Christians who claim that up to thirty thousand were burned at the stake during the Spanish Inquisition alone. The truth is often bracketed by the extremes. There is no question that tens of thousands were tortured by the Church in front of Church appointed Inquisitors and that many thousands were relaxed."

"Relaxed?" I asked.

"The term "relaxed" is the ultimate euphemism. It meant that the victim was turned over to civil authorities for burning at the stake."

Sipping my wine, I reflected. "As I recall, Christians were persecuted, crucified and fed to lions in their early history. The Inquisitions were created by their descendants?"

"Ignorance knows no bounds, Sarann. The Inquisition sought out Jews, Moslems, any budding sect of Christianity or virtually any other unofficial faith or lack of faith. There is no question in my mind that many of the fundamentalist sects that have budded off of the original faith would repeat the Inquisitions in every detail if they ever held enough power to do so."

"*NOBODY expects the Spanish Inquisition*!" a voice in my head suddenly blurted out. I ignored it. Too much wine brings them on sometimes.

"It must have cost a fortune to prosecute all of this on such a grand scale," I suggested.

"Actually, the Inquisitions generated a great deal of wealth for both the Church and the Crown. The possessions of the convicted were confiscated. After a few hundred years of Inquisitions, the Church and the governments of different nations within the sphere of the Vatican's influence ended up owning virtually everything of value."

"What finally brought it all to an end?" I asked.

"They petered out from a lack of financial incentive in the early 1800's. There was almost nothing left to plunder."

"I don't quite understand why anyone would admit to not being of the proper faith considering the ramifications of doing so."

"Well, some members of the Church eventually asked that question themselves. They realized that some people might make false admissions just to stop the torturer from crushing more of their toes, or ripping out their fingernails, or breaking their bones, or burning especially sensitive body parts. The best solution the Church could come up with was to ask the victim to confirm their admission after being tortured. Of course, if they changed their mind at this point they risked being tortured all over again."

"Not what you would call foolproof, eh?"

"Not hardly," Jesús said as he stood, signaling that it was time to retire to his study for further conversation.

Jesús poured us some coffee. I sat back and prepared to listen.

He began. "The words of the Bible were written many centuries ago. Mankind has learned a lot since then. We have learned that we are animals, animals with a capacity for deep thought, love, and empathy. What is it that sets humankind apart from the rest of God's creatures? It certainly isn't our physical appearance. All primates have fingers and toes, two eyes and a nose. We all eat and defecate. There is only one thing, just one thing. We are unique in our capacity for love and empathy. Not just for one another, but for God, and all of the other creatures we share this planet with. There is your answer.

"How could a book written in an ancient language thousands of years ago give us guidance on things like birth control, the Internet, amniocentesis, genetic engineering, abortion, paranoid delusional schizophrenia, ecology preservation, biodiversity, or extinction?"

Taking my cue, I asked. "I don't know… how?"

"It can't. The original authors were almost unimaginably ignorant by today's standards. The author's didn't know that a heart pumps blood or that the brain creates thought. They had no idea that plagues were caused by tiny creatures too small to see. Plagues became the wrath of God. The mentally ill were possessed by demons. They didn't know of the past existence of dinosaurs and mastodons. They did not know that the sun is a nearby star, one of billions in a galaxy that today is called 'the Milky Way' which is itself just one of billions of other galaxies each filled with billions of stars. They lived in a very small world. They were unaware of most other civilizations coexisting on other parts of the planet.

I interjected. "So, are you saying that in the context of modern societies, the words of the Bible are inadequate?"

"No. What I am saying is that the Bible is complete bullshit."

"Bullshit?" I said, not sure if he was joking. "What brought you to this conclusion?"

"A few moments of honest critical thought."

Here I was, sitting in the study of a brilliant theologian, being told that the Bible is bullshit. I wasn't sure how to come back to that. I decided to stall, "I have to confess, I really don't know any of the history of how the Bible came to be. Maybe you should give me a brief synopsis."

"Very well. The Bible is a translated composite of thousands of Hebrew and Greek manuscripts that were written hundreds of years after the death of Christ in the case of the New Testament and hundreds of years before the birth of Christ in the case of the Old Testament. None of these original manuscripts exist—only copies of copies of copies, passed along for hundreds of years and all copied by hand. Even though most scholars agree that the spoken language of Jesus was Aramaic, his words and deeds first came to us as ancient copies written in two different languages not his own."

"What are you getting at?" I asked.

"I have found irrefutable evidence that the Bible has an accumulated linguistic error rate that approaches 100 percent."

"I think you lost me there somewhere, Jesús. What does that mean?"

"Over time, the Bible has degenerated into pure nonsensical gibberish. I have suspected that something was wrong for quite a while. 'So God created man in his own image, in the image of God he created him; male and female he created them.' No one talks like this. It isn't old English; it isn't a dialect of English at all. No one has ever talked like that. Even the name of Jesus has been completely twisted with time.[81]

He stood up and walked over to a blackboard. "Yah-sh-ua (Hebrew) was translated to Le-s-ous (Greek) to le-so-us (Latin) and finally Je-s-us (English). The Hebrew alphabet doesn't even have a "J" sound in it. Look at the original name and the last one. They bear no resemblance to one another. This twisting has happened to every word in the Bible. What does it mean to have been created in God's image? Literally, it means that God looks like a naked, tailless, male primate. Is that what they meant to say?"

Caught slightly off guard, I stammered. "Uh, come to think of it, probably not."

He continued, "Actually, that is exactly what they meant to say. Those who were making copies of copies, actually thought that God looked like a human being, more specifically, a male human being. They had no idea that there were apes in Africa that look and act very much like us. Certainly, the concept of us being primates was lost to them. When they made their copies and translations, they interpreted what they saw as best they could."

"Many believe that the various translations and copies of the Bible are guided by the hand of God and are therefore immune to error," I said, just for the hell of it.

---

[81] http://www.plim.org/JesusOrigin.htm

Jesús gestured toward a wall lined with hundreds of Bibles. "Every Bible in this collection has multiple typos, misspellings and omissions. Collecting Bibles that vary from one another has become a hobby of mine. God is definitely not guiding any hands in this instance. The vagueness of the Bible leaves it open to interpretation. These interpretations vary from decade to decade and from sect to sect. This is an indisputable fact. Galileo spent his old age in house arrest because of a phrase in the Bible that said that the Earth does not move. He was lucky, one misstep and he would have been burned at the stake."

"Where is all this leading?" I asked, trying to get to the meat of it.

"It came to me as a revelation while doing a simple linguistic experiment in a language class that I was teaching. Using a child's game sometimes called 'operator,' or the 'whisper game,' I was trying to demonstrate how easily communication could be distorted. I asked my students to form a semi-circle. I whispered two short sentences into the ear of the first student who in-turn whispered them to the next student and so it went around the circle. The results of this exercise vary a great deal depending on how long or complex the initial phrases are and how many people participate."

"So what happened?" I prompted.

"Coincidentally, the resulting phrase whispered into my ear by the last student in the semi-circle was an exact quote from the Old Testament."

"And...?"

"The thought suddenly occurred to me to repeat the experiment starting with this quote. Would I get my two original sentences back?"

"Did you?" I asked with an air of expectancy.

"Well, no, but it was an interesting idea."

The raspberry sound signaling disappointment came unexpectedly from my lips.

"Sorry to disappoint you, but the science of linguistics just isn't that simple," Jesús declared defensively.

"So what did you do next?"

"Realizing that the Bible might just be the result of thousands of years of people inadvertently playing the operator game

I developed a computer program that could take the results of a game of operator as input and produce a list of probable phrases that had been used at the start of the game. The program uses a combination of sophisticated linguistics techniques, probability algorithms, and information theory to generate a list of possible solutions. I then began using quotes from the Bible as input, tracing it back through time from modern day English to ancient dialects of English to Latin to Koine Greek and on through to ancient Hebrew and from there to Aramaic."

"What did you find out?"

"I told you already."

"Oh, right, the Bible is bullshit—a lot of good that does you. So, no Heaven, no Hell, no disciples of Satan."

Smiling, he said, "Televangelists come as close as you'll get to being disciples of Satan."

I didn't know what to say at that point.

Jesús continued, "Armageddon is upon us. It will not take the form of a fiery battle with the destruction of the Earth in the balance. The battle has already been joined with the end of nature in the balance. You could say that we are in a struggle with Satan and his disciples and all of the millions of unfortunate souls who have succumbed to their false prophecies. Armageddon is the battle to save the biodiversity of our planet."

I wasn't sure what to make of all this. When in doubt, ask more questions, I always say. "So you're saying that the destruction of the planet's ecosystem is a kind of Armageddon in slow motion. You're rather isolated in the religious sector with this idea aren't you?"

"Actually not. One of the fastest growing sects of Christianity in the U.S. is Evangelical Environmentalism. I have tentative plans to join forces.[82]"

"So what are you planning to do with the output from your program?"

---

[82] Evangelicals are coming to the (earth's) rescue" by Dan'l C. Markham dated April 28, 1997 in the High Country News, http://www.hcn.org/servlets/hcn.Search?keywords=evangelicals&search=search

"I have condensed all that is worth saying into a handbook that is small enough to fit in your back pocket. It will be published as the 'Recovered Testament.' It is my hope that it will someday replace the Old and the New Testament versions of the Bible."

"You're publishing a new version of the Holy Bible?" I said, taken aback.

"That's correct. No vagaries, no misinterpretation."

"Whew, I'd love to take a look at this thing."

"I'll give you a copy before you leave. The computer code used to generate it, however, is sacred text," he said with a wink.

"Why are you compelled to start yet another religion? Couldn't you just pick one and run with it?" I asked feeling somewhat overwhelmed.

He turned to a file cabinet and pulled out a thick folder. "Here is an A-Z list of religions. Which one would *you* suggest?"

I looked at the first few pages, which just covered religions that begin with the letter A:[83]

Aaronic Order, Abecedarians, Abelites, Abenaki, Abenaki, Passamaquoddy, and Penobscot, Abipon, Abkhazian, Aboriginal Evangelical Missions, Abyssinian Frontiers Mission, Acharya's Yoga and Meditation Centre, Acheh, Acholi, Achomawi, Acolapissa and Tangipahoa, Acts Full Gospel Church, Adai, Adamites, Adja, Adonis, Aduma, Advent Christian Church,AdventistAetherius Society Afar affiliated, affiliated - black, affiliated - full communicant, Africa Evangelical Church, Africa Evangelical Fellowship, Africa Inland Church, African Baptist Assembly, Malawi, African indigenous churches, African Methodist Episcopal Church, African Methodist Episcopal Zion Church, African Orthodox Church African Orthodox Church of the West, African People's Socialist Party, African Protestant Church, African Reformed (NGK-South Africa), African Traditional Religion, African Union First Colored Methodist Protestant Church, African Universal Church, African Zion Church, Afro-American Army, Afro-Brazilian religions, Afro-

[83] The list of religions was obtained from http://www.adherents.com compliments of Reginald Finley (no relation).

Cuban religions, Agapemonites, Aggressive Christianity Missionary Training Corps, Aglipayan Church, Agni, Agnostic, Agon-shu, Agudat Israel World Organization, Agul, Ahanta, Ahirs, Ahl-I Haqq, Ahmadiyya, Ahmadiyya – mosques, Ainu, Ais, Aiyetoro Community, Aizo, Ajivikas, Aka, Akan, Akposo, Alacaluf, Alamo Christian Foundation, Alar, Alaska Yearly Meeting, Alaskan Athabaskans, Alawi, Albakourat al-Durzeyat, Albani-Sinai Orthodox, Albanian, Albanian Orthodox, Albanian Orthodox Archdiocese in America, Albanian Orthodox Church in America, Albanian Orthodox Diocese of America, Albigensianism, Aleut, Alevis, Alexandrian Wicca, Algard Wicca, Algemene Doopsgezinde Societeit, Algonquin, Alianza Evangelica Menonita, All Africa Conference of Churches, Allegheny Wesleyan Methodist Connection, Alliklik, Alpha and Omega Pentecostal Church of God of America, Alsacian Churches, Alsea, Altays, Alternative Religions, Altkolonier Mennonitengemeinde, Altkolonier Mennonitengemeinde (Colonia Manitoba), Altkolonier Mennonitengemeinde (Colonia Nueva Durango), Altkolonier Mennonitengemeinde (Colonia Rio Verde), Aluk To Dolo, Amahuacas, Amal, Amalgamated Flying Saucers Club, Amana Church Society, amaNazaretha, Ambo, Ambonese, American Association of Atheists, American Association of Lutheran Churches, American Baptist Association, American Baptist Churches in the U.S.A., American Bible Society - volunteers, American Carpatho-Russian Orthodox Greek Catholic Church, American Catholic Church, American Catholic Church (Syro-Antiochean), American Council of Christian Churches, American Eastern Orthodox Church, American Episcopal Church, American Ethical Union, American Evangelical Christian Churches, American Evangelistic Association, American Family Association, American Gospel Band, American Holy Orthodox Catholic Eastern Church, American Independent Orthodox Church (Bridges), American Indian Evangelical Church, American Lutheran Church, American Lutheran Conference, American Muslim Mission, American Muslim Society, American National Baptist Convention, American Nazi Party, American Orthodox Catholic Church (Irene), American Orthodox Catholic Church (Propheta), American Orthodox Catholic Church, Archdiocese of Ohio, American

Orthodox Catholic Church, Western Rite Mission, Diocese of New York, American Orthodox Church, American Prelature, American Rescue Workers, American Unitarian Association, American World Patriarchs, Americanism, Americans United for Separation of Church and State, Amhara, Amidism, Amish, Amish - other, Amway, Ana, Anaak, Anabaptist, Ananaikyo, Ananda Community, Ananda Marga, Ananites, ancestor veneration, Anchor Bay Evangelistic Association, Ancient and Accepted Scottish Rite, Ancient and Mystical Order of the Rosae Crucis, Ancient Church of the East, Ancient Tridentine Catholic Church, Andamanese, Andes Evangelical Mission, Andhra Evangelical Lutheran Church, Andhras, Angkola and Mandailing, Anglican, Anglican - active, Anglican - attend at least yearly, Anglican - confirmed, Anglican Catholic Church, Anglican Church of Canada, Anglican Church of North America, Anglican Communion, Anglican Episcopal Church of North America, Anglican Mission in America, Anglican Orthodox Church, Anglican Rite Jurisdiction of the Americas, animism, Anthroposophical Society, Anthroposophical Society - full-time, Anti-cult movement, anti-missionary movement, anti-Semitism, Antigua Barbuda Baptist Association, Antiochian Orthodox, Antiochian Orthodox Christian Archdiocese of North America, Apa Tanis, Apache, Apalachee, Apayao, Apinaye, Apostelamt Jesu Christi, Apostolic, Apostolic - historic Apostolic succession, Apostolic - independent, Apostolic - other, Apostolic Assemblies of Christ, Apostolic Catholic Church of the Americas, Apostolic Christian Church, Apostolic Christian Church (Nazarean), Apostolic Christian Churches of America, Apostolic Church, Apostolic Church (Australia), Apostolic Church of Christ, Apostolic Church of Christ in God, Apostolic Church of Jesus, Apostolic Church of Jesus Christ, Apostolic Church of Pentecost of Canada, Apostolic Church of Queensland, Apostolic Faith, Apostolic Faith (Hawaii), Apostolic Faith (Kansas), Apostolic Faith Mission, Apostolic Faith Mission Church of God, Apostolic Faith Mission of Portland, Oregon, Apostolic Faith Mission of South Africa, Apostolic Gospel Church of Jesus Christ, Apostolic Lutheran Church of America, Apostolic Lutheran Churches, Apostolic Lutherans (Church of the First Born), Apostolic Lutherans (Evangelicals No.), Apostolic Lutherans

(Evangelicals No.), Apostolic Lutherans (Heidmans), Apostolic Lutherans (New Awakening), Apostolic Methodist Church, Apostolic Overcoming Holy Church of God, Apostolic Spiritual Baptists, Apostolic United Brethren, Aquarian, Aquarian Foundation, Aquarian School of Yoga, Aquarian Spiritualist Centre, Aquarian Tabernacle Church, Ar nDraiocht Fein, Arab, Arab & Muslim combined, Arab Liberation Front, Arachne, Arapaho, Arapaho and Atsina, Arapesh, Araucanians, Arawak, Arbeitsgemeinschaft Mennonitischer Brudergemeinden in Deutschland, Arbeitsgemeinschaft Mennonitischer Gemeinden in Deutschland, Arbeitsgemeinschaft zur geistlichen Unterstutzung in Mennonitengemeinden, Arcane School, Arianism, Arica, Arikara, Aristotelian, Armenian, Armenian Apostolic Church, Armenian Apostolic Church of America, Armenian Apostolic Church of America - Eastern Prelacy, Armenian Apostolic Orthodox Church of America, Armenian Church of America, Armenian Church of America, Diocese of the, Armenian Evangelical Church, Armenian Orthodox Church, Arminianism, Armitage Baptist Church, Arya Samaj, Aryan, Aryan Brotherhood, Aryan Nations, Aryan Warriors, Asatru, Asbury Bible Churches, Ashaninka, Ashanti, Ashkenazi Judaism, Ashurism, Asmat, Asociacion Bautista de El Salvador, Asociacion Convencion Bautista de Costa Rica, Asociacion Convencion de Iglesias Menonitas de Costa Rica, Asociacion de Iglesias Hermanos Menonitas de Colombia, Assam Baptist Convention, Assassins, Assembleias de Deus, Assemblies of God, Assemblies of the Lord Jesus Christ, Assembly of Brothers, Assiniboine, Associacao das Igrejas Menonitas do Brasil, Associacao dos Irmaos Menonitas de Portugal, Associacao Evangelica Menonita, Associate Reformed Presbyterian Church (General Synod), Associated Brotherhood of Christians, Associated Churches of Christ, Associated Gospel Churches of Canada, Association des Eglises Evangeliques Mennonites de France, Association for Research and Enlightenment, Association Mennonite Luxembourgeoise, Association of Baptist Churches, Association of Baptist Churches in Israel, Assyrian, Assyrian Church of the East, Assyrian Church of the East (Chaldean-Syrian/Daly), Assyrian Evangelical, Assyrian Orthodox, astrology, astrology - astrologers, astrology - full-time astrologers, astrology - part-time astrologers, Atakapa, Athanasianism, Athapaskans, Atheism, Athiopisch-Orthodoxe Kirche, Atlantean, Augsburgian Confession, Augustana Evangelical Lutheran Church, Auldearne Witches, Aum Shinrikyo, Aum Shinrikyo Auroville, Australasian Conference of Seventh Day Baptists, Australian Aboriginal religion, Australian Aborigines, Australian Conference of Evangelical Mennonites, Australian School of Yoga, Australian Transmission Meditation Network, Autocephalous Slavonic Orthodox Catholic Church (In Exile), Autocephalous Syro-Chaldean Church of North America, Avar, Avoyel, Aymara, Azali Babis, Azande, Azerbaijani, Aztec.

"I see your point." I said.

Jesús continued, "Robert and I have been corresponding. I have taken some of his advice to heart and have designed this religion to provide my followers with what they need. Is there anything I've told you tonight that rings false?"

"No. This doozy is airtight all right."

I called Bob shortly after returning home from my visit with Jesús.

"Bob, I just got back from Mexico City. Were you aware that Jesús is publishing a new version of the Bible?"

"Well, no, I haven't been privy to that information. It doesn't surprise me though."

"I'm a little concerned about his sanity," I added.

"I wouldn't worry about Jesús. His definition of God is the absence of evil. When you find goodness, you've found God. His definition of Satan is the polar opposite. Satan is the absence of goodness. You have found Satan when you find evil. All you have to do now is define good and evil."

"But, isn't that the hard part?" I asked.

"True. Just the same, I believe he's saner than the two of us put together."

I couldn't argue with that one.

After talking with Bob, I had a vague remembrance of an E-mail he had sent me a few years ago. I dug through my computer files and found it in a folder labeled miscellaneous. It was a transcript of an Internet chat conversation between himself and Jesús:

Dear Sarann

I have sent you this Internet transcript with Jesús' permission. You asked us to keep you informed. I suspected from the start that his religion needed some spice. If he accepts my advice, he will have a distinct advantage over his competitors. They have buried the truth in their subconscious minds. Jesús will have the truth at the conscious level. He also won't be limited to ancient, dated dogma that contradicts itself at every turn. He can make up his own.

Sincerely,
Bob

Jesús: "Robert, I am seeking advice."

Bob: "Shoot."

Jesús: "My ratings are not what they should be. I cannot put my finger on it."

Bob: "Well, Jesús I was afraid this was coming. What your followers need is consolation that they will never die—that there is an afterlife waiting for them—and that your God is willing to physically intervene on behalf of the group who prays the hardest."

Jesús: "But that's not the truth."

Bob: "No shit."

Jesús: "Are you suggesting that my religion is flawed?"

Bob: "No. I'm suggesting that you stretch the truth just a little."

Jesús: "You can't be serious."

Bob: "Oh but I am."

Jesús: "I do not feel comfortable with that statement."

Bob: "Imagine that its 1942, you are in Amsterdam, and a little girl by the name of Anne Frank lives in the attic above you. A squad of Nazi storm troopers just asked you if you know the whereabouts of a girl by that same name. What would you have done?"

Jesús: "I would have lied."

Bob: "To save the life of one little girl?"

Jesús: "Yes. I wasn't really expecting this."

Bob: "Get on with it, Jesús. The planet needs you. We all need you. Clear, rational, logical, defensible thought has very little to do with organized religion. You're gonna have to shine 'em on, boy. Do you really think that any televangelist completely believes the crap they spout? They don't. On a certain level of consciousness, they have their nagging doubts. They have to work very hard to keep the truth hidden from themselves and by extension, from their followers. That is why they go to such extremes to reinforce their beliefs through ritual and conversion. That is why they surround themselves with others who all want to believe the same thing, never questioning, never accepting debate, preaching to the choir, attacking non-believers as though they were the enemy. Envision rows and rows of big-headed primates with their hands clamped firmly over their ears and eyes but not their mouths. Fight fire with fire, Jesús."

Jesús: "This comes as a shock to me."

Bob: "I'll tell you something else. You have to come up with an equivalent for the Koran, Torah, or Bible. If you don't get your dogma down in print you aren't going to get anywhere. Mark my words."

Jesús: "I have to think about what you have said."

Bob: "I wouldn't expect anything less. Good luck to you."

Personally, I was relieved that Jesús had not taken all of Bob's advice… or had he?

CHAPTER 17

The roast

I was in New Orleans for a face-to-face think tank meeting. I prefer virtual meetings, but on occasion we will get together in a real room around a real table. Bob's cigars were always a problem, Dave could not employ his time delay profanity software, and diffusing Teresa's propensity to talk without end took a lot more diplomacy. Jesús on the other hand was a pleasure to have around regardless. His charismatic personality lent an air of tolerance and cooperation to the gatherings.

We were all meeting for lunch. The actual tank meeting was scheduled for later in the afternoon. Bob and I had rooms at the Bourbon Hotel in the French quarter. I knocked on his hotel door.

"Come in!"

The smell of cigar smoke was overpowering. "Phew! This is supposed to be a non-smoking room. Let's go, I'm starving."

"Have a seat," Bob said to my reflection in the mirror as he finished shaving.

I sat down in an overstuffed chair.

"You know, Sarann, I have to admit, you were not what I was expecting when I walked into your office for my job interview all those years ago."

"Oh really? Why does that not surprise me?"

"I deserved that."

"We all stereotype, Bob. It is a natural and necessary aspect of our human nature. Like so many other aspects of our nature, it evolved to benefit the individual, often at the expense of others. If it makes you feel any better, I forgive you."[84]

---

[84] The instinctive nature of stereotyping is discussed in Chapter 8 of Volume II.

He walked past with a towel over his shoulder. "I know. Just the same, who would have predicted that I would one day have as my best friend and confidant someone like you?"

"Gee, how does one come back to a compliment like that?"

Smiling, he said, "Well, I've had my limit of touchy-feely bullshit for one day. Let's roll." He walked past me and into the hall.

I pulled the door shut and ran to catch up. As we entered the lobby, I thought I saw a familiar face getting into a cab. I had trouble placing it at first, but then it came to me. "Bob, I think I just saw one of Pinwright's lawyers get into a cab."

"Have you been taking your medications, Sarann?"

"As a matter of fact I have." I retorted, unable to shake the feeling that something was up.

We decided to walk rather than take a cab. As we waited at a red light, I looked to my left. "Ben Pinwright? What brings you to New Orleans?"

Startled to see me, he stammered, "S-Sarann? My, well, yes... "

Bob interrupted, "You must be here for a conference or something."

"Yes! Yes, that is why I'm here. Good to see you both," he said as he walked off waving.

I turned to Bob. "Something is going on. That guy can't tell a lie to save his soul."

"I hear you. Let's keep our eyes open," he said as he donned a pair of dark sunglasses.

Great… now I had piqued Bob's paranoia.

We arrived at the restaurant. Our group was to be seated outside. I saw Jesús, Teresa, and Dave waving to us through a door that led to the patio. They were already seated. As we entered the courtyard, I saw that it was full of people.

A surprise party. I love surprise parties. The air filled with applause and cheers. After a little handshaking and backslapping, I took my seat as the guest of honor. Everybody was there, and I mean everybody. I saw Oscar and Jason from the Colombian contract, Leto from the Congo contract, Chekov from the Siberian, Marcy from the Indiana hike, the old lawyer, Ben, Randy, his wife, and many others.

Bob took the podium first. "Well, Sarann, you have suffered my second-hand smoke and jokes for many years. I find it ironic that thanks to John Nash, I cannot say you're the first schizophrenic ever nominated for a Nobel Prize.[85] At least I can say that you are the first African American schizophrenic. Congratulations, you know I love ya."

Bob had made progress through the years. When I first met him, he was still using the word Negro.

Next up on the podium was Teresa. Her wheel chair silently rolled up to the podium and turned to face the crowd. "Sarann, you taught me a great many things as I watched you lead this think tank of misfits through all of these years. It was because of you I met my husband Dave and now have a wonderful daughter. I'm not very good at telling jokes so instead I have put together a slide show."

She had collected highlights from years of virtual tank meetings. The slides included clips of Dave falling off his chair, each of us in various states of head bobbing consciousness as we struggled to stay awake at meetings, and much more. The highlight—for me at least—was a clip of Jesús caught with a finger up his nose.

Much to my surprise, Chekov from the Siberian contract made his way to the podium. Having been the butt of one of his jokes, I just happened to know that Chekov speaks very little English. While in Siberia, I once had a long and enjoyable conversation with him only to find out that he understood not a word I said.

"I thought he doesn't speak English," I said to Bob.

"He doesn't. I wrote that speech for him in the Russian alphabet. He has no Idea what he is about to say."

"Oh."

Chekov began in his thick Russian accent: "Sarann, I am lower than dog doo. Scrape me from the soul of your shoe..." The more he read, the louder the laughter. Finally, he stopped, realizing that he was on the losing end of a joke this time. He turned, wagged his finger at Bob, and said, "I get you bastard!"

---

[85] A beautiful mind by Sylvia Nasar

Bob leaned over and whispered in my ear, "You're even now."

The roast continued on this way for another hour or so. Finally, Jesús stood and approached the podium.

He began. "I first met Sarann—also known to some of you as Uhura, Pat, or Chris—twenty years ago at a job interview, as did the other members of our renowned think tank. Her goal was to save the planet. She may have succeeded. The world's population will soon begin a slow decline and most experts believe that enough critical habitat is being protected at this point to hold off further extinction while our population continues to shrink to a more viable level.

"Born and raised in a small one-room garage with no indoor plumbing, Sarann was the third child in a family of six children. Neither of her parents could read or write and that is how the unique spelling of Sara Anne came to be. Sara Anne, it is my great pleasure to announce that you have achieved Sainthood in the Church of the Recovered Testament as the savior of our planet. I raise my glass to you. For the Earth!"[86]

[86] I hope you were not offended just because you didn't see this ending coming. It was meant to drive home my point that there is a thing called human nature and that it is not always a nice thing. It usually manifests itself at the subconscious level and therefore cannot consistently be overridden with intellectual effort.

CHAPTER 18

Epilogue

"Sarann, over here, these bear tracks look fresh."

We were in the Sealth preserve, the largest temperate forest in North America. Grizzly bears, wolves, and mountain lions are the top predators here, not man. The rivers and lakes are teeming with fish. The entire area is completely uninhabited by humans and ringed with the largest fence on Earth. The fence is not to keep people out, but to keep the wildlife in. No one wants what is inside this wilderness.

Our primary mission was to census wildlife. We were heading toward a spiraling column of buzzards to investigate what they were feeding on. Bison had proliferated and because of that, California condors had finally returned to the wild. The Condors had lived on the carcasses of bison for many thousands of years and were doing so once again. We were standing next to a stream packed with spawning salmon. There was a fishing ban on all pacific coast salmon. Bald eagles sat in the trees and along the bank feasting on the fish. We continued to the small lake that was being fed by the stream. A moose stood among the numerous beaver lodges. We were safe in our suits but prudence is always called for in this kind of habitat. Grizzlies in this preserve have lost all fear of people.

The year is 2040. Thanks primarily to my namesake, the world population stopped far short of 9 billion and has been gently declining for some time now. Her vision of continuing urbanization had materialized and the mass extinction has been defused. The march of technology has continued unabated and hunger has been eliminated. Of course, people continue to fight, hate, and argue without end. Violent conflicts come and go with the usual regularity, but at least we are getting a handle on how to protect our biodiversity. Uncle Bob's security organization—EPI—is alive and well to this day.

AIDS finally mutated itself into oblivion. A benign version began to circulate that prevented infection from the deadly version. A decade later, the scourge was gone.

The transition from fossil fuels to hydrogen is almost complete. Back in 2004, it looked like hydrogen had no hope of replacing oil because, unlike coal and oil, there are no vast reserves of it just waiting to be collected and disseminated. Hydrogen has to be created first and doing that took more energy than you got back. The Holy Grail of energy production has always been a fusion reactor. Surprisingly, we still have not perfected that technology. All energy today comes from fission reactors. They generate the electricity that converts water into hydrogen. The key to safely using fission came in the form of new technology that allowed us to neutralize radioactive waste. Decommissioned reactors are now torn down and completely deradiated. All radioactive waste is treated on-sight thus giving us fission reactors that do not generate radioactive materials that need to be transported or stored. As far as safety is concerned, the same frictionless, superconducting containment fields that make my modern Sentinel suit impervious to everything from dust to grizzly bears also protect nuclear plants. The probability of meltdowns, leaks, and fires is so low now as to be essentially impossible.

Today, a Sentinel suit looks like a highly reflective, person-shaped mirror. The suit seems to disappear while standing in jungle foliage because it reflects everything around it.

Advances with superconductors created a surge in novel technologies. People in 2004 thought technology was growing rapidly but they had no idea. A power plant in Los Angeles can send its surplus electricity all the way to New York without any loses. Power plants can now operate at peak efficiency twenty-four hours a day. Because there are negligible transmission loses in the wires, fewer plants are necessary, and when they are not meeting peak requirements during the day, they are making hydrogen at night.

The machines that transport us today make the old combustion engine cars look comically inefficient. A spider had been found in Central America that has a web five times stronger than any other. Scaled up, the synthetic version of this web was used to make materials that are 25 times stronger than steel. This was a breakthrough for flywheel technology. The problem with storing energy in a flywheel is that you need tremendous rotational velocities. The problem with these high rotation rates is that they put extreme stresses on the flywheels. When a flywheel breaks, it is like a bomb going off. The spider web allowed engineers to build super strong flywheels and an encasement that could contain one should it fail. Now our vehicles get their energy from flywheels that turn at ungodly velocities inside vacuums on frictionless bearings. The energy from the flywheels can be converted into electricity or kinetic energy. The flywheels are kept up to speed with a tiny ceramic turbine engine that, of course, burns hydrogen.

The Sealth preserve owes its existence to the Chernobyl effect. The radioactive fallout from the Chernobyl incident had acted as a species-specific biocide. Long-lived, slow breeding species could not exist there anymore—people for example. Birth defects and sky-high cancer rates kept them out. However, high cancer rates and a few birth defects have little impact on wild boars, wolves, and deer. They do not live long enough for cancer to affect their populations. In addition, their birth rates are high enough to overcome stillbirths caused by deformities. As a result of having no human beings in the area the Chernobyl dead zone had quickly become rich in wildlife.[87]

On September 1, 2007, it was announced on the evening news that a terrorist organization may have hidden a nuclear device near Seattle. The president made a special appearance to explain that the CIA had reliable and detailed intelligence confirming this as an elaborate hoax meant to cause panic and disruption. In a reassuring tone, the citizens of Seattle had been told to stay calm and go about their daily business. Public confidence in CIA intelligence and presidential assurances being at an all-time low, the exodus had started immediately. Two days later Seattle was a ghost town.

The National Guard was called out to search for the weapon that did not exist but might be detonated at any time. Under-

---

[87] http://www.kiddofspeed.com/chapter1.html

standably, the desertion rate exceeded 90 percent. However, the effort paid off. The bomb was discovered and defused on September 10th. In a move to instill confidence, and to get the citizens of Seattle back to their jobs, the president flew to Seattle on September 11th, landing on an aircraft carrier in Puget Sound. In a live broadcast, with the Seattle skyline in the background, he assured the world that America could not be cowed by acts of terrorism. As the president continued to talk, a second window popped up with a live camera shot of a huge and growing mushroom cloud. A thermonuclear device had been detonated in Seattle's harbor. The terrorists had planted a second bomb just in case. Ah, human nature. Fortunately or maybe unfortunately for the president, the entire aircraft landing had been staged the day before. It hadn't been live after all.

It was eventually learned that these bombs had been bought for ten million dollars apiece from the Russian Mafia. Russian authorities had been aware of the thefts but chose to keep the information quiet, confident that they would be able to find them before it was too late. Following the law of unintended consequences, Iraqi terrorists fueled by instinctive predilections toward hatred and revenge had obtained the weapons using funds provided by a deeply religious and wealthy individual who had been told by his God to destroy evil. Yet another dull-witted and religiously devout heir to oil money had made a huge mess of things after consulting with his higher father.

People have not changed, but at least the sixth mass extinction has been halted.

Index for Volume I

Prologue II

The rise of hominids triggered a protracted and relentless decline in the ability of all other species to compete for living space. Humans, consciously or not, drive extinctions.

Wild creatures must have contiguous ranges in which they're free to move about, procreate and find alternate food sources.

Biodiversity forms an interconnected system incorporating all animals, plant, insects and microbes. Evolution has unified and refined it over billions of years. Attempts at redesign of the natural biosphere always have unintended consequences, mostly bad.

The preservation of biodiversity and our own short-term and long-term interests are inseparably linked because any aspect of robust natural variety, once extinct, is gone forever.

Because of human nature, overpopulation is trumping conservation and will continue to drive biodiversity loss. We must continue to seek lower fertility rates and even allow our numbers to shrink someday in the future. In the meantime, we can apply a tourniquet by accelerating ecosystem conservation. [88]

What can you as an individual do? Well, more than you might think. You can give on an annual basis to a conservation organization *and* to a women's reproductive rights organization. The more you can spare the better. Money is a convenient form of power and it makes a *significant* difference. The Nature Conservancy and Conservation International are making great headway in protecting critical ecosystems. Without the availability of safe pregnancy termination procedures, what little is left would already be gone. You will see what I mean as you read on into volume II.

---

[88] Jeremy M. Harris, one of the editors for this book, wrote the preceding five paragraphs as part of a critique. I used them here with his permission.

Volume II

This portion of the book can be read independently from Volume I. It is a collection of essays on the subjects of nature, human nature, evolutionary biology, and the synthesis between them that has put in motion the sixth great extinction event in the history of our planet. They can be read in any order. Keep in mind that up-to-date Internet links for the footnotes can be found at the website www.poisondarts.net.

Contents Volume II

CHAPTER 1

The TIFIC

The TIFIC—take it and forget it contraceptive—is a long-lived contraceptive system for men and women consisting of a pill or capsule that causes infertility until another pill or capsule reinstates it. Any contraceptive concept that does not meet this definition in its entirety fails to earn the label of TIFIC.

Researchers have been working on anti-fertility vaccines for about three decades now. In 1995, a petition signed by over 430 groups and organizations from 39 countries called for an end to research on contraceptive vaccines—the TIFIC, by the way, does not have to be in the form of a vaccine.[89] The petition had its roots in the international feminist health movement. In response, the World Health Organization (WHO) publicly stated that they would halt research if the potential users said they did not want this kind of contraception. The petition was prompted by fears of government control over the reproductive rights of women.

I can empathize with anyone concerned over a stumbling government bureaucracy with a long-lived contraceptive vaccine at its disposal. That is precisely why the TIFIC must be at least temporarily reversible. However, I do not agree with any group that tries to protect itself from a potential harm in the future by stifling research. A TIFIC can never become a reality without further extensive laboratory effort. All research has the potential to uncover technology that can be abused by power brokers, but that is no excuse to crush the pursuit of knowledge. The longev-

---

[89] Judith Richter: Vaccination against Pregnancy: Miracle or Menace? (Zed Books, London, 1996)

ity of a contraceptive is only a matter of degree; some contraceptives last longer than others and some are more effective than others. The TIFIC is not a euphemism for chemical sterilization.

Unintended pregnancies cause untold hardship. This is especially so in third world countries where the depth of poverty can be almost unimaginable. Even here in the U.S. the effects of a "take it and forget it contraceptive" (TIFIC) would be unprecedented. The number of single mothers on welfare would plummet. The abortion debate would evaporate. Teen pregnancies would be reduced to rare anachronisms.

Ironically, the TIFIC may become a reality because of infertility research. Free market forces are driven by the quest for profit. Infertile couples routinely pay tens of thousands of dollars for work-ups. When was the last time anyone spent that kind of money on a contraceptive? One team of researchers stumbled upon an idea for a contraceptive while investigating infertility in women who are allergic to sperm. They discovered how to duplicate this immunity to sperm, and the idea of using it as a contraceptive was born along with a patent. In fact, I first uncovered this information while doing a patent search for contraceptive vaccines.

Vaccination research is thriving.[90] There are several contraceptive schemes afoot.[91] One candidate to become the TIFIC involves a newly discovered ion channel protein.[92] This protein is located in the sperm flagellum and is found nowhere else in the human body. Its job is to control the flow of calcium into the tails of sperm. Calcium is the trigger that causes the tail to beat. If a vaccine antibody is developed to block this ion channel, it could be taken by both sexes.

One vaccine contraceptive has been used on 95 species of animals to date. The vaccine contains a protein called procine zona pellucida (pZP) that is obtained from pig eggs.[93] When injected, it stimulates an immune response creating antibodies that attach to the ovulated egg and prevent sperm from attaching to the egg.

---

[90] http://www.niaid.nih.gov/vrc/
[91] http://www.aafp.org/afp/20040215/853.html
[92] http://www.hhmi.org/news/clapham2.html
[93] http://www.ovpr.uga.edu/researchnews/spring98/elephant.html

In some instances, the animals are vaccinated using bait that is laced with the vaccine. In other instances, the animals are injected. This technique is only about 90 percent effective and therefore not acceptable as a human contraceptive. It also works only on females.

In one project a population of foxes was brought under control only to have several of their prey species overpopulate. Ironically, a vaccine program was then considered to control those out of balance populations. People never cease to amaze me.

Controlling wild animal populations in this manner has its drawbacks. One is that some animals will exhibit immunity to the vaccine. Natural selection will rapidly cause the entire population to become immune. This problem wouldn't apply to people. Finding immune foxes would be difficult. Tests will be available to identify immune people. For immune foxes, the elimination of their competition would help to fuel their population increase. That would not be the case with people. In theory, people who are immune to the TIFIC would continue to have unplanned pregnancies and, *given enough time*, would populate the world with their offspring. This would take a very long time and the rapid growth of technology should be able to outrun that bullet quite easily. In addition, by having a male and female version, the odds that both members of a pair bond would be immune would be very low.

A TIFIC raises many concerns. For example, attempts to control the populations of rival ethnic, religious, or political groups will be made. You can't change human nature. I do not see that as a serious threat or, at least, it is not a new one. An effective antidote would circumvent that possibility. No government would be able to crush a black market trading in antidote fertility pills. Witness how ineffective our own government's war on drugs has proven to be.

The TIFIC could be used to take reproductive rights away from women. Again, the antidote would negate that concern. A version for men would also help to defuse that possibility.

There is the concern that descendants of those who choose not to take a TIFIC will overrun the world. That would take a very long time. In addition, there would be nothing keeping their descendants from taking a TIFIC just because their parents chose not to. This is not a realistic concern.

For various physiological reasons, not everyone will be able to take a TIFIC. That's OK, this idea doesn't require universal compliance. Taking a TIFIC is voluntary. The motive for taking it is to dodge unplanned pregnancies. The fact that it will help save the planet's biodiversity is actually a side effect.

There is a concern that sexually transmitted diseases will increase with the introduction of a TIFIC. This would require an increase in sex. The TIFIC will not have that effect. Sex is a very complex social act. Young men who set out to end their virginity soon find out just how difficult that can be.

When you think about it, the idea for a vaccine to protect people from sperm isn't far-fetched at all. Look at the vaccines that already exist—smallpox, rabies, polio, tetanus, and anthrax to name just a few. There are vaccines against poisons from scorpions, spiders, and snakes. There are hundreds upon hundreds of vaccines with hundreds more on the way.

Now that the human genome has been unraveled, we are poised for another exponential technology explosion in the direction of genetic engineering. This may be how we finally get the TIFIC.

One of the many interesting details exposed by the genome project is that retrovirus infections have had an impact on our genetic code.[94] This has been suspected since the discovery of retroviruses. Retroviruses attack certain cells in the body and integrate part of their own genetic code into that of the infected cell. This is how HIV works. It becomes a part of the target cell, turning that cell into a virus-making machine. Unfortunately the target cells are critical to our immune system and once their genetic code is modified they no longer act as originally designed; that is why HIV causes disease. If a retrovirus were to infect sperm cells or egg cells, the inserted piece of genetic code would pass from generation to generation as with other genetic diseases such as cystic fibrosis. The genome record shows that this has happened many times in the past. Part of our genetic code has come from retrovirus infections that spliced their DNA in with ours.

---

[94] Nature Via Nurture by Matt Ridley, page 114.

The implications of this are fascinating. For example, will we find that retrovirus infections of sperm and egg cells have been a driving force in evolution? Could they account for episodes of extremely rapid evolutionary change within a species? Did retrovirus infections create the social structure seen today in termites, ants, bees, and wasps? What if HIV had a propensity to infect human egg cells instead of cells that are critical to our immune system? Instead of HIV being a deadly disease resulting in a destroyed immune system, it would cause infertility. Envision such a disease, but make it as easy to catch as the common cold. It wouldn't take long for most people on the planet to be exposed. Women who were immune to the virus because of genetic variability would be the only ones able to procreate.

This would create the human equivalent of queen bees. Would human culture suddenly lurch in the direction of the social insects? A better analogy would be the reproductive strategy of naked mole rats. As with the social insects, only the queen mole rat reproduces. The fact that they are highly social, hairless, mammals, makes the analogy that much more unsettling. I am digressing again but this idea would make a nice backdrop for a science fiction novel.

Using genetic engineering to modify viruses—particularly retroviruses—is a potentially dangerous exercise. It has to be done with safeguards in place, using biohazard protocols and equipment. One such experiment for a mouse contraceptive in Australia accidentally turned a relatively benign mouse-specific virus into a deadly one.[95]

What I would like to see is something akin to a Nobel Prize awarded for the development of a TIFIC—something that would bestow great status, prestige, and wealth on the inventors. If the prize were large enough, unprecedented—fifty million dollars or so—we would have the TIFIC within a decade.

The key to the profitability of this contraceptive and it's antidote is not so much in its selling price, but in the fact that so many people would use it—6.35 billion customers and more on the way everyday.

Picture walking into a pharmacy. Sitting on a shelf are boxes labeled TIFIC. Each box comes with a single capsule and a

[95] http://news.bbc.co.uk/1/hi/health/1110144.stm

simple urine test kit. The instruction sheet inside the box warns you not to engage in unprotected sex for five days after taking the medication. At that time you must also dip a test strip into a urine sample to make sure the TIFIC is active in your blood. The instructions might say that your partner should also use protection to decrease the chances of a contraceptive failure. The instructions could also suggest—for marketing purposes—that the best choice for your partner's contraception is the TIFIC.

If I had millions of dollars to invest, and if I were looking to go down in history as the person who saved the biodiversity of the planet, I would pour it into the development of the TIFIC. The sooner it happens the better. If the U.S. has an unintended pregnancy rate that exceeds 50 percent, imagine what it must be in other parts of the world. The first company to market this product will blow all other manufacturers of contraceptives out of the water. My message to the researchers, entrepreneurs, and industrialists of the world—build the TIFIC, there is money to be made.

CHAPTER 2

Overpopulation—the modern denial

The word overpopulation was brought into the modern public consciousness in 1968 by a book called *The Population Bomb* written by Paul Ehrlich. The book was a warning to the world that the human population was a ticking time bomb about to explode. The book badly missed the mark with its predictions of resource shortages and famine. For example, Ehrlich predicted that four billion people would starve to death in the 1980s. Based on that prediction, he also called for some highly unpopular measures to reduce our population growth. To ice the cake, he made a bet with an economist named Julian Simon about the price of five metals over a ten-year period. Ehrlich said the price would go up as a result of dwindling resources and Simon said the price would drop. The price dropped. Ehrlich paid up. Why Ehrlich thought that new sources for those metals would not be found or that manufacturers would not make compromises to compensate for shortages is beyond me. As an engineer at Boeing, I often altered a design to use high strength aluminum instead of titanium because the cost of titanium was too high. At one time, most sources of titanium were in the Soviet Union. We used titanium extensively after the fall of the Berlin wall.

I recently met Paul Ehrlich at a signing for his latest book *One with Nineveh.* I listened to him speak for an hour. He is an impressive figure, tall, vigorous, with a deep voice and a tremendous sense of humor. I was very impressed.

Anyone who makes the mistake of predicting human trends with precision is asking for trouble. One can predict with relative certainty that the stock market will crash or boom, but you would be rash to say when those things will happen or to what extent. People with a propensity for betting are going to lose once in a while. Simon's betting record wasn't flawless. In 1997 he paid

$1000 to David South, a professor at Auburn University, over the predicted price of pine sawtimber.[96] That's the problem with gambling, you have to stop when you're ahead.

A later book, *The Population Explosion*, published in 1990 by Ehrlich and his wife, Anne, did a much better job documenting the havoc wreaked upon the planet by our huge numbers. The title of the book suggests that the population bomb has gone off, and evidence of that blast can be seen everywhere.

The population growth advocates point out that because of improved technology, food production presently outstrips population growth. They also believe that resource shortages will always be compensated for with new technologies. However, they are making the same mistake that the Ehrlichs made. They are extrapolating what exists today into the future.

In 1968, the same year that *The Population Bomb* was published, with a third of the world hungry, and with the percentage of world population growth maxed-out at about 2.1 percent, it looked like Ehrlich's concerns were dead on. The Ehrlich's had assumed that because a third of the planet was already hungry we had reached the limits of our ability to feed ourselves. Logically, this left only one option, reducing population growth. Coincidentally, 1968 was also the last year to see an increase in the *percentage* of world growth—not *total* growth—and was the start of a decline in *percentage* of growth that continues to this day—see Figure 1. However, predictions of extremely rapid population growth held true and Ehrlich's predictions of mass starvation would also have panned out if our food supply had not unexpectedly increased even more rapidly than our population.

Although the percentage of population growth began to decrease in 1968, the actual number of people being added every year continued to increase as a result of population momentum reaching a peak of about 87 million in 1989—see Figure 1. Again, by coincidence, this was the same year that the *Population Explosion* was published and also marked the start of a decline in the number of people added every year that continues to this day. Approximately 75 million people were added to the world population in 2003.

---

[96] http://www.forestry.auburn.edu/sfnmc/web/bet.html

Year	Population	Growth	# added per year
1964	3,276,816,764	2.08%	69,021,089
1965	3,345,837,853	2.08%	70,227,393
1966	3,416,065,246	2.02%	69,742,104
1967	3,485,807,350	2.04%	71,868,340
1968	3,557,675,690	2.08%(Bomb)	74,665,661
1969	3,632,341,351	2.05%	75,268,761
1970	3,707,610,112	2.07%	77,580,647
1971	3,785,190,759	2.01%	77,006,527
1972	3,862,197,286	1.96%	76,511,302
1973	3,938,708,588	1.91%	75,889,828
1974	4,014,598,416	1.82%	73,625,631
1975	4,088,224,047	1.75%	72,167,756
1976	4,160,391,803	1.73%	72,536,792
1977	4,232,928,595	1.70%	72,474,692
1978	4,305,403,287	1.74%	75,373,540
1979	4,380,776,827	1.72%	75,928,390
1980	4,456,705,217	1.70%	76,259,715
1981	4,532,964,932	1.76%	80,436,954
1982	4,613,401,886	1.73%	80,530,264
1983	4,693,932,150	1.68%	79,634,655
1984	4,773,566,805	1.68%	81,036,085
1985	4,854,602,890	1.70%	83,004,818
1986	4,937,607,708	1.73%	85,962,468
1987	5,023,570,176	1.71%	86,583,085
1988	5,110,153,261	1.67%	86,179,948
1989	5,196,333,209	1.67%	87,422,136(Explosion)
1990	5,283,755,345	1.56%	83,182,744
1991	5,366,938,089	1.53%	82,725,730
1992	5,449,663,819	1.48%	81,337,993
1993	5,531,001,812	1.44%	79,976,536
1994	5,610,978,348	1.41%	79,887,428
1995	5,690,865,776	1.36%	77,746,508
1996	5,768,612,284	1.35%	78,192,518
1997	5,846,804,802	1.32%	77,770,099
1998	5,924,574,901	1.31%	77,934,526
1999	6,002,509,427	1.29%	77,632,256
2000	6,080,141,683	1.26%	77,258,877
2001	6,157,400,560	1.24%	76,849,827
2002	6,234,250,387	1.22%	76,299,210
2003	6,310,549,597	1.19%	75,477,418
2004	6,386,027,015	1.16%	74,526,549

Figure 1: Population Data
Source: U.S. Bureau of the Census, International Database.

Following is an example of how the percentage of growth can fall but the total population can continue to grow. Assume you have 10,000 mice and their population grew by 10 percent in a mouse year. That would give you 1,000 more mice for a total population of 11,000 mice (0.1 x 10,000=1,000). The next mouse year the population increased only 9.9 percent adding 1089 mice (0.099 x 11,000=1,089). You can see that 89 more mice were added the following year than the previous because 9.9 percent of 11,000 is a bigger number than 10 percent of 10,000. The percentage of growth today is about 1.2 percent—see Figure 1.

Ehrlich's predictions missed their marks for several reasons. Keep in mind that everything I am about to say is in hindsight. Many things have come to light in the last 36 years that explain why the predictions never panned out. They were not part of the debate in 1968.

First, it has long been assumed that human population grows exponentially. To this day, if you search the Internet for the definition of exponential growth the authors will invariably use population growth as an example. Anyone who has had a course in biology knows the age-old example of lily pads on a pond. The most elegant version of this I have ever heard came from E.O. Wilson in his book *Consilience*:

> "A lily pad is placed in a pond. Each day thereafter the pad and then all of its descendants double. On the thirtieth day the pond is covered completely by lily pads, which can grow no more. On which day was the pond half full and half empty? The twenty-ninth day."

When each pad on the twenty-ninth day replicated itself, collectively, they covered the rest of the pond. In this case the analogy is an attempt to describe the rate at which humanity is destroying the planet's biodiversity, not the usual description of human population growth.

I am going to try to explain this without—as the young bucks I play basketball with would say— "getting mathematical on your ass." The most common example of exponential growth that almost everybody understands is an interest bearing savings account. If you plot the growth of your account over time you will eventually see the "j" shaped curve characteristic of exponential growth. However, if you closely inspect the shape of the actual human population growth curve—as opposed to a theoretical one—you will discover that it does not look like it was generated by exponential growth. It actually looks quite linear in most areas although it has had a very steep slope—has been increasing very rapidly—over the past 40 years.[97]

Ehrlich along with everybody else in 1968 assumed that our growth was exponential when in reality it was just growing extremely fast, which is a fine point and not really relevant unless you are trying to win an argument with a mathematician. In addition, it was thought that our food supply would grow linearly. Linear growth will always be outstripped by exponential growth, given enough time. What everyone missed is the fact that our technology growth is exponential and our food supply is a function of that technology.

Technology growth is exponential because technology begets more technology. For example, the ability to make iron allowed us to make iron picks and shovels which allowed us to mine more iron ore, make iron plows, internal combustion engines, and the steel oil rigs that extract the fossil fuels that run those engines and provide us with most of our fertilizer and plastics used to make computers… and on it goes. Remember, this has all come to light in hindsight. Nobody was presenting all of these ideas as a counter argument in 1968. Most everybody feared that the Ehrlichs were making a good point. The data at hand was looking very grim.

The final straw was falling fertility rates. People in the developed nations were having fewer children. As is typical with economists and stock analysts, long-winded explanations for why things happened begin showing up *after* they have happened. These explanations have all been collected under a heading called the demographic transition theory.

The online encyclopedia Wikipedia defines the demographic transition as follows:

> "In demographics, the term demographic transition is used to describe the transition from high birth rates and death rates to low birth and death rates that occurs as

---

[97] http://en.wikipedia.org/wiki/Malthusian_catastrophe. Also see How Many People can the Earth Support? By Cohen, pages 81-84.

part of the economic development of a country from a pre-industrial to a post-industrial economy."

Unlike most theories, no single person gets credit for this one. It was formulated and reformulated by groups of economists starting in the late 1920s and is still being continually revised.[98] Some don't see it as a theory so much as a collection of reasons that do a reasonable job of explaining why fertility rates have been falling.

Today's wealthy western countries went through a transition. Nutrition and medical science were improved by an ever-increasing technological database, causing a dramatic decrease in death rates, which in turn created a population explosion. This is exactly what you see happening in third world countries today because the green revolution combined with medical science has lowered child mortality. But, you can't just feed poor people and provide them with vaccines and antibiotics. You must also provide them with an education, family planning, and decent jobs. A decent job can also mean a well-run farm.

What happened next in the West was a drop in fertility. Fertility rates in some nations are below replacement level. Fertility rates are also dropping now in third world nations and the hope is that they too will make the "demographic transition" to low fertility rates as the western nations did.

The theory is anything but airtight. Exceptions abound around the world. Our own birth rates plummeted during the hard times of the great depression but went ballistic during the hot economy of the forties and fifties creating the baby boom generation. Both trends were counter to what the theory would have predicted.

There are, of course, arguments over what caused the world population to explode but most arguments are over what caused our fertility rates to fall and whether or not the third world will make the full transition.

Once again, keep in mind that no one wrote a best selling book in 1968 countering the Ehrlichs that said, "Relax everybody, just relax, fertility rates will fall and food production will then catch up. Giant well-funded NGOs like Conservation Inter-

---

[98] http://www.ub.rug.nl/eldoc/dis/rw/b.j.de.bruijn/thesis.pdf

national will come into existence to try to save what is left of the planet's biodiversity. Everything will be fine, really. Go back to sleep."

Following is an abbreviated list of reasons given for the falling fertility rates in the West:

1) People got jobs in cities and no longer needed children as indentured labor for farm work (urbanization).
2) People no longer had extra children as a hedge against high infant mortality thanks to better health care.
3) Women were emancipated; they became better educated and joined the work force.
4) Contraceptive technology improved.
5) Social security programs replaced the need to have children around to help out in old age.
6) Somehow the idea got planted into people's heads that it might be wise to have fewer children. Having fewer children became a fad. This implies that if a fad comes along in the future saying that having more children is the "in" thing to do, then we are going to see a fertility spike. For example, fertility rates for new immigrants from Mexico remain high because they are unaware that large families are not "in." They have not realized yet that their new peers in their new culture frown upon large families. Their children will not keep those same fertility rates, being more aware of their cultural norms.

Every last one of these points are contested or promoted by one group or another and each group has studies and reams of data to back them up. Why? Because we are human beings, and as human beings it is part of our nature to argue, and fight, and bicker at the drop of a hat, like Siamese fighting fish.

These six reasons are interrelated. For example, they all depend on reason 4—new and improved contraceptive technology. But, contraceptive technology is only good for avoiding unwanted pregnancies. Distributing contraceptives, providing sterilization and abortions does little good if people still want to have large families. Inversely, people who want smaller families need family planning to accomplish that.

Reason 2—reduced infant mortality—does not decrease fertility by itself. Think about it. Reduced infant mortality means increased fertility. Reduced infant mortality has to be combined with most of the other reasons for it to result in lowering fertility rates—fewer babies.

Logically, this suggests that further improvements in health care without doing the other things will just make more impoverished babies. In other words, you pretty much have to fund all of these things at once to have success. Improving the health care of desperately poor people without also getting them out of poverty, educating and empowering their women, providing family planning (the spacing of children, safe pregnancy terminations, contraceptives, and instructions for using them) will not reduce fertility rates. The commercials you see asking for money to feed hungry children and to provide them with health care may be making more of those hungry children. If that money were to simultaneously go to things that would reduce their poverty—educating and empowering their mothers and providing family planning—then you would have something. So, donate your money, but donate it in a manner that covers all bases. The road to hell is paved with good intentions.

Following is another abbreviated list, this time, of mindsets you will encounter in the population debate:

1) There are those who feel that poverty reduction is all that matters. Some groups promote the reduction of poverty over family planning. It is true that if you can reduce the level of poverty sufficiently, people stop going hungry regardless of family size. America's baby boom is an example of this, but for that to happen you have to have a very high standard of living. Fertility rates become unimportant with regards to hunger if you can raise the standard of living high enough and since fertility rates have had a tendency to fall with increased standards of living, why emphasize family planning? With this reasoning, all efforts should go to poverty reduction; family planning is largely irrelevant.
2) There are those who feel that it is not feasible to reduce poverty without a population reduction because we are con-

suming the planet in a non-sustainable manner already and poverty reduction will accelerate that consumption.

3) There are those who feel we will all learn to live sustainable lives and preserve the planet regardless of population size.
4) There are those who feel population reduction is all important and that the rate of reduction has to be greatly accelerated by increasing our death rates as well as reducing our birth rates.
5) Finally, there are those who believe none of the above. Neither population reduction nor sustainable lifestyles are necessary because human ingenuity and free markets will always prevail; the more people we have on the planet, the better off we are.

I will start by discounting mindset four—we must increase our death rates. I will not discuss why because there is near consensus in the world that it should be discounted and defending my position is not necessary.

The problem with mindset one is that family planning and poverty reduction are inextricably interlinked. The key is that you need family planning to reduce fertility rates; it is the means by which you accomplish a reduction in birth rates once women decide they want it. Poverty reduction is the mechanism that persuades women to have fewer children; family planning provides them with the means to accomplish it. They must be promoted simultaneously. For poor nations, a reduction in fertility is also necessary for a reduction in poverty—unless the extra children are helping to reduce poverty, as is the situation with child labor and subsistence farming. In which case, women are free to have more children anyway. However, in most instances, a poor woman with eight young children has no chance of pulling herself up and out of the poverty trap. They argue that programs that provide *only* contraception, sterilization, and pregnancy termination procedures are ineffective in reducing poverty. They are partially right of course, as I said earlier. But, a stool with only two legs is worthless. If you are not providing family planning at the same time you are striving to reduce poverty and protect the environment you have a two-legged stool. The reductions of poverty and population growth go hand in hand. Those who lobby that energy and money must be concentrated on poverty reduction and empowerment at the expense of family planning are ignoring the fact that there are about 40 million abortions performed annually. Take a minute and think about that.

Mindset two best describes my own, and is the most common mindset among those who strive to reduce population growth.

Mindset three, sustainable living, has potential to alleviate damage to the environment, however, the slogan, "live simply so that others may simply live" is flawed because, contrary to popular belief, my lifestyle here in America has little, if any impact on the life of someone in Africa. There is no mechanism to shift resources I do not use to those who need them more than I do. By eating less I do not provide more food for someone else. Selling my car will not help someone else live. As a voluntary movement, it has negligible impact. Asking those who have accumulated significant wealth and status to give it up is not going have much success. Profit motive and status seeking overrides most ideologies that do not have those drives as part of them. Surely, driving a small hybrid car will reduce emissions, but among environmentalists there is also a great deal of status associated with driving a hybrid car. Sustainable living alone will not be enough with nine billion people struggling to improve their lives: the idea is overwhelmed by the numbers. There is not enough iron on the planet to provide nine billion people with a hybrid car.

Mindset five—the more people we have the better off we will be—was championed by a conservative economist who I will introduce to you shortly.

The Ehrlichs were major players in the worldwide social fad, or meme, that put the idea into billions of people's heads that it was wise to limit the size of their families—reason six. The word meme was coined by Richard Dawkins in his book *The Selfish Gene*. He defined the meme as a unit of intellectual or cultural information that survives long enough to be recognized as such, and which can pass from mind to mind.[99] The meme idea is quickly becoming a meme in its own right. It is probable that the

---

[99] http://maxwell.lucifer.com/virus/alt.memetics/what.is.html

Ehrlichs were instrumental in starting the overpopulation meme, and that this same meme may be one of the reasons why the world population is predicted to peak between eight and ten billion instead of twelve to fourteen billion. The Ehrlichs certainly were not the first to discuss the idea of overpopulation. Malthus and others had written extensively on it. But the Ehrlichs were the ones most responsible for disseminating the idea to the common man through their popular books. The work done by the Ehrlichs is one reason why we don't have a much bigger population today or the famines they predicted.

Predicting the future often changes the future, thus nullifying the prediction—especially when trying to predict human behavior. I experienced an example of this paradox just yesterday. I was supposed to pick a friend up at the airport. The previous evening our TV news media had predicted dire traffic congestion because one of our two north-south highways was going to be closed for repairs. In addition, fog was expected which was going to make things even worse. Seattle is notorious for its traffic jams.

Ignoring the warning from these *doomsayers*, I set out to retrieve my friend in the middle of rush hour and set a new speed record for getting to the airport because the highway was practically deserted. The *pessimistic* prediction of horrific traffic snarls had kept everyone but the most foolhardy off the roads. Because they never materialized, should I have thumbed my nose at those who predicted traffic snarls or should I have thanked them?

It is highly probable that the work done by the Ehrlichs has had the same future-altering effect on slowing the devastation of our planet as well as improving the plight of humanity.

Had the percentage of population growth remained at 1968 levels through 2004, there would be about 2 billion more people on the planet right now. Instead of having 800 million hungry today, we might have 2.8 billion hungry. There is no doubt that the warning from the Ehrlichs was a factor in lowering fertility rates. It also helped light the fires that created today's gigantic and well-funded relief organizations. Should we be thumbing our noses at the Ehrlichs or thanking them?

As a side note, if all abortions had been prevented throughout the world—as our current administration would like to see—

our population might be about 12 billion today. It took about five minutes to build a spreadsheet to generate these numbers. You can build one for yourself to check me. I used 2.1 as the percent of growth starting with a population of 3.4 billion in 1965. There were about 40 million abortions performed worldwide last year—roughly 50 percent of the number of people added—so I assumed that without abortions, population growth every year would have been increased by about 50 percent.

Countering the Ehrlichs was the late Julian Simon. His books, *The Ultimate Resource* and *The Ultimate Resource 2* were embraced by economic conservatives, the Catholic Church, and pro-life groups. The books were also used to buoy policy by both Ronald Reagan and Pope John Paul II.[100] Simon urged the Reagan administration to cut off of funding to the International Planned Parenthood Federation. This led to the curtailment of U.S. support for worldwide family planning that continues to this day under the Bush administration.[101]

Simon believed that the more people we have on the planet, the better off *people* will be. He felt that with more people, there would be more geniuses like Einstein. He understandably made no mention of the greater number of Stalins and Pol Pots we would also have.[102]

One of the weakest links in Simon's works is that the preservation of the planet's biodiversity does not play a big role in the plan. The closest he came to addressing that issue was by showing how the first world countries have cleaned up their air and most of their rivers and lakes. You can swim in them again, and rivers no longer catch on fire. However, those victories belong to the antithesis of Julian Simon, the environmental activists who successfully enacted legislation, not to the industrialists now burdened with the cost of cleaning their effluents. Protecting the environment rarely has any potential for short-term profit. It usually costs money to protect the environment and therefore

---

[100] http://www.cato.org/pubs/policy_report/cpr-20n2-1.html

[101] http://www.goodbyemag.com/jan98/simon.html

[102] A review of Simon's book by Herman E. Daly can be found at http://www.mnforsustain.org/daly_h_simon_ultimate_resource_review.htm. This review was originally published in the Bulletin of the Atomic Scientists, January 1982.

legislation has to be forced down the throats of those who lose profit as a result of that legislation. Our water is cleaner, but most sturgeon and salmon runs are gone and many freshwater bivalves are endangered or extinct. Introduced species dominate the fauna (carp) as well as the flora (milfoil). The biodiversity that once existed in our lakes and rivers continues to degrade.

A major contribution Simon made was in highlighting the fact that people are very adept at seeking solutions to their problems, finding ways to compensate for resource and food shortages. That is why we invented farming and domesticated animals.

The unknowns that bit the Ehrlichs are waiting to bite Simon's work as well. The interplay between dwindling natural resources, global economies, climate fluctuations, incurable plagues, a huge human population, and most of all, human nature, is far more complex than the simple idea "the more, the merrier." Simon did not publish his book until 1980. This was twelve years after the publication of *The Population Bomb*. Simon looked at the census data and put what he saw into a book. Food production was outpacing population growth and economies were growing. The predicted shortages had not materialized. He was counting beans. His work consists of reams of data documenting what was happening. The data eased a lot of minds, his in particular and provided fuel for conservative economists and pro-life groups.

I believe that Simon's greatest contribution was in documenting the awesome power that properly regulated free economies have to create incentives for profit that in turn create new technology and make it affordable to the common man *even in the face of an exploding human population*. The magnitude of this exponential growth in technology was totally unanticipated by the Ehrlichs. However, it isn't necessary to have 6 billion people alive simultaneously to have exponential technology growth. We already had automobiles, airliners, antibiotics, vaccines, nuclear bombs, skyscrapers, television, and toaster ovens in 1945 when our population was about two billion. The planet was not a lonely desolate place then. Two billion is still a huge number. Technology growth was fully exponential. We have managed to stay fed and housed as a result of technological and economic growth, *not population growth.* It is not a coincidence that technology and population have both been growing rapidly. There is some correlation, in that medical and agricultural technologies have decreased infant mortality, and these are the overwhelming reason for our population growth. However, our huge population is not causing our technology growth. The lion's share of new technology comes from the industrialized countries, some of which have stable or even declining populations. The people in India, China, and Africa, have certainly benefited from these innovations and technologies, but the existence of all these people was not critical to the creation of these same technologies. The computer, the green revolution, and the Internet would have happened if the populations of India, Asia, and Africa had not grown since 1945. In fact, with fewer people living in squalor, there may have been even more contributions to technology from those parts of the world because the standard of living would have been higher.

The world's reserves of iron were estimated at 20 billion metric tons in 1950. They were estimated at 93 billion metric tons in 1980. A similar scenario exists for oil and many other resources.[103] This is the primary reason why the prices of these commodities remain low. The fact that we have managed to discover more natural resources is hardly evidence that we will never run out of them. Simon said that the only way for sure to know if resources are getting scarce is to look for rising prices, and since prices are not rising, there are no scarcities. What goes unsaid is that most of the prices are not rising because we keep finding reserves—oil, and iron for example—that we did not know existed or that we did not know how to get at. It also means that we have not carved up the last of the forests or crashed the last of the fisheries. For example, we make fake crab meat out of fish that we used to consider trash. That is not the same as saying that we are not using up our resources. That only says that we have found ways to buy more time. Simon's view was that we will always find ways to buy more time. We will always be able to make more from less.

---

[103]http://www.davidrhenderson.com/articles/0698_inmemoriamjuliansimon.html

I would prefer not to take the risk. People want control over the timing of births, why not give them the means? A world stripped of its biodiversity by nine billion people is also a scenario we need to avoid. At the time of this writing, the number of undernourished people had started growing again by about five million a year.[104] Place your bets everyone.

I am not going to devote this book to a point by point critique of all of Simon's views. Suffice it to say that his views were not overly concerned with the biodiversity of the planet. In Simon's view, if a species cannot prove itself to be valuable to people, then it has no value. Simon once said,

> "We do not neglect the die off of the passenger pigeon and other species that may be valuable to us. But we note that extinction of species billions of them ... has been a biological fact of life throughout the ages, just as has been the development of new species, some of which may be more valuable to humans than extinguished species whose niches they fill."[105]

This gives insight into Simon's world perspective and personally, it sends a chill down my spine. He simply was not concerned about the fact that our zoos are filling with the last specimens of various species like the Newly deer that have become extinct in the wild. When you read an article on the Internet supporting Simon's views, you will also find that they are universally devoid of any mention of the planet's loss of biodiversity. Interestingly, you will also find that people with highly conservative or religionist viewpoints wrote the articles.

The truth is that *some* of Simon's views are correct and so are *some* of Ehrlich's. The solution is a combination of the ideas of both men—a reduced population and a healthy world economy with free trade generating technology that has the potential to save us *and the planet's biodiversity.*

---

[104] http://www.news24.com/News24/World/News/0,,2-10-1462_1450334,00.html

[105] (Simon and Kahn, The Resourceful Earth: A Response to Global 2000, P. 23)

Ultimately, Simon's view that things only get better with a larger population has at least four fatal logical flaws. First, if the answer to the world's problems is to increase our population—suggesting that 40 billion is preferable to 9 billion—what then is the upper limit? Is there an optimal number and if so, what is it? Is it 40 billion, or 10 billion, or was it 2 billion?

Next, the idea that you will have more people like Einstein if you have more people, is nullified by the fact that you will also have more people like Stalin, Hitler, Pol Pot, Mao, Idi Amin, and Saddam Hussein (who does not really rank with the others). You will also have more serial murderers, poachers, psychopaths, thieves and rapists. When someone like Darwin or Einstein contributes to our knowledge database, their contributions stay with us. They are additive, each contribution building on the next over the centuries. It is not necessary that they all be living at the same time to contribute. The more people you have, the more bad guys you have, and they contribute nothing to our future database. They screw things up for finite periods. The theory of evolution and the theory of relativity would all have gotten here eventually if our population had never bloated to what it is today. In fact, they may have gotten here sooner if we had a smaller, richer, and better-educated population.

Third, the free market forces that continue to find ways to feed and house us will always try to do so at the expense of the environment if that is the way to maximize profit, which is usually the case. Legislation enacted by environmentally minded people is the only thing that stands in its way.

Finally, Simon's view that humanity will always find innovative ways to compensate for resource shortages plays right into the hands of people like me trying to find a way to save the planet's biodiversity. If humankind will always find alternatives to compensate for shortages of things like copper and lumber, then there is no reason not to rope off all of the remaining ecosystems and get on with the business of finding creative alternatives. In other words, if Simon is correct—and when combined with Ehrlich's idea of a sustainable population size he just might be partially correct—then there is no reason *not* to rope off all of these places as an insurance policy for future generations. Why don't we leave the old growth forests alone? Why are we cutting down the Congo for its hardwood? Why are we cutting the Amazon down to feed cows? We will quickly adjust and find alternatives shortly after the commodities are made unavailable to us. Although it was not his intent, Simon has given the conservationists of the world just what they need to save it. Thank you Mr. Simon, may you rest in peace. Maybe, someday, the liberals and conservatives will override their instincts to form into self-righteous hate groups and give you and the Ehrlichs the respect you all deserve.

Many of the problems of the world are too complex for specialists. Paul Ehrlich, being a biologist, is poorly equipped to deal with the economic side of the equation and Simon, being an economist, was poorly equipped to deal with the biological side. This concept of pulling together multiple disciplines and sciences in an attempt to solve the world's problems is outlined in E. O. Wilson's book *Consilience: The Unity of Knowledge*. It turns out that most issues are interrelated. A multi-discipline approach is the only way to tackle them.

Late in the game (1995) a book was finally written called, *How Many People Can the Earth Support?* by Joel E. Cohen.[106] He had the advantage of hindsight having read the works of Ehrlich and Simon. Cohen took every precaution not to fall into the same trap as Ehrlich. He did this by refusing to predict the future. By refusing to be pinned down, he managed to write a book on overpopulation that was almost controversy free. This proved to be the book's weakest link, leaving many readers wondering why they had bothered to read this 500-page tome. You can't change the future if you do not have the bravado to predict it. Intrigued by the title, many readers picked up the book hoping to find the answer. The answer, although correct—that it all depends—leaves us unfulfilled.

Cohen starts by making light of the numerous examples all through history of people who were concerned about overpopulation. The earliest example going back 4000 years. People have always harped about overpopulation. In other words, "the sky is falling" story has been around for a long time. This is an exam-

---

[106] A review of this book can be found at http://www.kzpg.com/Lib/Pages/Books/96-04-12__How_Many_People_Can_..Cohen.html

ple of why population growth advocates often quote from the book to support their beliefs that overpopulation is a fabrication. What Cohen failed to convey, or possibly failed to see, is that overpopulation has always been a local problem. These people lamenting about overpopulation 4000 years ago had very legitimate concerns. Their small worlds *were* overpopulated. When the people of Easter Island were starving to death, the planet was far from overpopulated. The archaeological record is rife with examples of populations that have crashed because of overpopulation, leaving behind parts of the planet where the environment has been so damaged it has never recovered. I recommend that you read *Constant Battles* by Steven A. LeBlanc if you are interested in how commonly this has occurred throughout ancient history.

The people of Easter Island starved although there was plenty of food on the planet. There just wasn't enough food where they lived—in their world. They were trapped by the ocean that surrounded them. Similar barriers, usually political and cultural, constrain all people. These barriers are just as real as the ocean and just as dangerous to cross. They have always existed and always will. When an impoverished villager walks to another village only to find more impoverished villagers, he has a problem. It is only when the numbers of impoverished villages begin to grow to encompass entire countries that you will find a global population crisis, which is exactly where we are today.

Another aspect of Cohen's book that disappointed me was the dearth of discussion on the biodiversity of our planet. Out of 532 pages he devotes just four of them to this topic. Admittedly, they are four very well written and powerful pages. In fact, I recommend that you read them next time you find yourself in a library. The sub-heading is *Time Constraints in Biological Diversity*, pages 336 through 340.

No one could predict an end to population growth until the percentage of growth stopped increasing. Thanks to falling fertility rates, demographers can finally predict roughly when our population will peak, and roughly what our population will be when it does—depending on assumptions of fertility and death rates. As we get closer and closer to the peak, it will get easier and easier to predict.

The fertility rates in some countries have already dropped below replacement level and have populations that are about to decline. There have been several pieces done in the media discussing this phenomenon. I read an article in the New York Times and saw a PBS documentary on the subject. In both cases, the media did a poor job describing to the public just what is going on. Many people who read the Times article or saw the documentary walked away with the impression that the world population was decreasing. I have seen it happen on two separate occasions. The individual misled by the Times article was a physician and the individual misled by the PBS documentary was an engineer. No wonder the public is confused on the issue. I have also heard three separate guests on National Public Radio say that the world population is decreasing. Nothing could be further from the truth.

Setting the record straight, we can expect our population to increase about 50 percent in the next fifty-to-eighty years, peaking at eight to ten billion. One billion people were added between 1988 and 2000. At the time of this writing, there are 6.35 billion of us.[107]

If you type overpopulation into a search engine, you will be confronted by wildly varying opinions on the subject. I just did so and the following is a brief synopsis of what I found. Of the ten sites on the first page, only half were concerned with the negative effects of a burgeoning human population. One of the sites is hosted by a pro-life, ex-catholic, atheist and promotes the belief that the planet is *not* overpopulated. He argues that technology will always come up with ways to stretch our resources. It is a direct reflection of Julian Simon's work. According to this site, air and water are getting cleaner, food supplies are growing larger, poverty is being reduced, and all of these positive trends will not only continue into the future, but will get even better with a growing population. Another site claims that we have passed our sustainable limits and the world population is crashing as a result of disease and decreased fertility rates. It predicts that our population will peak at about 6.9 billion in about twenty-

[107] http://www.un.org/esa/population/publications/wpp2002/WPP2002-HIGHLIGHTSrev1.PDF
(click on highlights and go to page 6)

five years and continue to drop because we have destroyed our life-sustaining ecosystem. The population having already passed through 6.35 billion, I think they may have to update their site any day now. A member of the Catholic hierarchy in Seattle hosts another site and jovially confuses the term overcrowded with overpopulated. Then come the sites that talk of pet overpopulation. Finding the wheat amidst all of the chaff is no easy task, and I wish luck to you.

The recent slowing of growth is a victory of sorts for population reduction advocates. Their efforts over the last three decades have had a much bigger impact than most people realize, although it may prove inadequate to save our biodiversity. Those who taunt the family planning activists by pointing out that humanity's population is going to peak at numbers lower than those previously predicted, are inadvertently highlighting the success of those who have fought so hard to curb our growth. Once our population has peaked, the new battle will be to allow our numbers to decline. Family planning advocates will have an easier time of it because momentum, contraceptive technology, and most importantly, human nature—the desire to time births for personal quality-of-life issues—will be on their side. The simple fact that humanity can for the first time predict roughly when our population will finally peak is the shining light of hope in thirty years of effort to slow our population growth—if we can just continue to avoid famine while simultaneously protecting the planet's biodiversity through it all.

There are those who believe we can protect the planet's biodiversity even with nine billion humans striving to maintain and improve their lives; there are even more people who could care less that we protect it at all. It is a collective case of self-deception to say we can protect an area in perpetuity by calling it a preserve. Government instability in places such as the Congo or Indonesia will be the final death knell for gorillas and orangutans. Experience with my own piece of private forestland has taught me much. It is remote and accessible only through rough logging roads, yet it is trampled by people. Trespassers leave their trash and toilet paper everywhere in addition to cutting tree branches and picking brush. Tree poachers, hunters, campers, dirt bikers, and four wheelers all leave their marks. A time-lapse camera would show a relentless flow of humanity over this landscape. History is filled with examples of wealthy individuals who owned nature preserves—usually for their own recreation. Sherwood Forest of Robin Hood fame is one of them. They all fall in the end to the pressures of population growth.

One of the demographic characteristics of a population that declines because of low birthrates is an aging trend. Before our population begins to drop, we will find that our average age will be much higher than in the past. The only way to avoid that scenario would be to increase death rates instead of decrease birthrates. This most certainly is not an acceptable option. At some point the world will have to grapple with a population that is top heavy with the elderly. I am going to say something now and repeat it a short time later for emphasis: Dealing with this aging issue is a moot point as far as the TIFIC is concerned because that day is coming with or without the TIFIC. The TIFIC would only move that day of reckoning up. The aging of our population will be a temporary condition because as we older people go off to heaven and hell, the natural and healthy ratio of young to old will eventually reach equilibrium again. This process will take some time, and cultural adjustments will have to be made. Again, it is a moot point as far as the TIFIC is concerned because that day is coming with or without it.

I can only guess what those adjustments would be. Resources that are now being spent to care for children may be shifted to care for the elderly since there will be no more unplanned pregnancies and therefore far fewer mothers needing welfare. Will an aging world population decrease the risk of war, allowing us to safely shift the trillions being spent on military budgets to other things? Would it be better to make the necessary adjustments at a lower population level or wait until our population peaks?

One statistic we may see as the population ages is a declining crime rate. Crime rates often correlate with the number of young men in a population. Young bucks will be less numerous. The costs of controlling crime may drop. An aging population will create a demand for things that young people do best. Young men will be too busy making money to be out causing trouble.

Another concern with a shrinking population is that it may cause economic hardship. There are many that believe you cannot have economic growth without population growth. If this statement is true, there is a day of reckoning approaching since population cannot grow forever. Again, would it be better to address the issue when there are nine billion people? An economist is yet to be born who could predict an economic trend in advance. Economists make a living explaining in detail why trends happened.

It's entertaining, at least to me, to watch the stock analysts on the nightly business reports explain why the market did whatever it did. It's like watching a show on astrology or palm reading. Most people don't seem to care that analysts never successfully predict what the market will do ahead of time. That would be like asking a weatherman to predict the weather for a month in advance. It is suspected that both the weather and the world economy emulate chaos theory. The modern world has never experienced a decreasing population. We really don't know what will happen to economies. The last person to ask is an economist. They are historians, not soothsayers.

The knowledge we have to date pertaining to economics was gleaned in a world with an increasing population. Our economy may actually continue to grow in relation to the number of people because that number will be declining, resulting in a stable or improved standard of living.

Anything I say here, and anything an economist may say on the same subject is little more than speculation. I can pick up a newspaper and read about the economic problems created by our huge population and then read an article one week later telling me about the economic problems that will be created by a decreasing population. That is just one reason why I don't read newspapers. Their articles are, out of necessity, too brief to thoroughly discuss topics of any complexity.

Expect to see groups starting to promote pregnancies as our population growth slows further. They call themselves pronatalists.[108] Will the development and distribution of the TIFIC become the next battleground for the pro-life and pro-choice activists? I'll bet my right arm on it.

---

[108] http://www.unfpa.org/swp/1998/newsfeature1.htm

France—along with Japan and some other countries—is attempting to grow its own workforce by appealing to patriotism and providing tax incentives for citizens to have more children. They see this as preferable to letting immigrants in to share the wealth. In 1920, France went so far as to outlaw all forms of contraception because so many young men had died in WWI. On the other hand, the governments of Korea, the Philippines, India, and China are working hard to reduce fertility rates.

When you stand back and look at our behavior, it all seems somewhat primitive: slaughtering our young men in massive wars, then cajoling our women to have more babies, or not to have more, depending on which country you are talking about. What I see are animal instincts of group protectiveness overriding rational thought in modern global economies. I see human nature at work.

There has been a huge mass migration across the planet from rural lands to cities in the last few hundred years. In 1800 only 3 percent of the world's population lived in cities and towns. By 2030 almost 70 percent of the world population will be living in cities.[109] This tendency will probably continue with a shrinking population. For example; the overall population of Mexico may decline while the population of its large cities continues to swell. This trend may continue all over the world. The economies of the cities may continue to be fueled by migrating workers while rural areas depopulate. If you ask why people are migrating to cities, you will get different answers depending on which economist you are talking to. In a nutshell, people are moving to cities because they perceive that they will be better off there. Why they have this perception can be debated, but perceptions change reality as often as reality changes perceptions—a wag the dog scenario. This fact—that human behavior cannot be described with equations—is the bane of all economic theorists.

The economic equations of supply and demand only work in a growing economy with a growing population. With a shrinking population, those equations will be turned on their ears. For example, with fewer people there will be less demand for lumber. Normally this would create a glut on the market and lower

---

[109]http://www.prb.org/Content/NavigationMenu/PRB/Educators/Human_Population/Urbanization2/Patterns_of_World_Urbanization1.htm

prices. The lower prices would mean less profit for the sawmills and layoffs for workers at the sawmills. With a shrinking population, however, you have fewer people needing jobs at those sawmills and fewer sawmills. A shrinking population does not automatically mean a shrinking economy. Population is a variable in the chaotic equations of economics.

Here is what I envision: As the world population shrinks, the existing trend of mass migration toward large cities to find employment and an improved quality of life will continue. At first this will tend to depopulate rural areas, rather than cities. This migration will be a source of labor fueling city economies much the same way immigration has fueled the U.S. economy. As the population continues to shrink, many cities will eventually begin to depopulate. These citizens will migrate to other cities with stronger economies. I see humankind continuing to lump together into cities linked through free trade. This will require countries like France and Japan to change their nationalistic attitudes and open their borders—in a controlled and logical manner—to immigrants the way the U.S. has *historically* done instead of prodding their citizens to have more French and Japanese babies.

Eventually the world population will stabilize with a natural and healthy ratio of young to old. The exponential growth of technology would continue—along with economic productivity—because technology is less dependent on population size than on free market forces and the fact that technology begets more technology.

Paradise would not be the word I'd choose to describe the world when we had a population of two billion considering that fifty million people had just died in the aftermath of World War II.[110] The baby boom in America was about to go full-tilt, signaling the start of a world population explosion that is presently adding 75 million people a year. The passenger pigeon, Tasmanian tiger, Dodo, and countless other species were already long gone. The American Bison had come to within a hair's width of extinction, almost following the Auroch into oblivion. We can do tremendous things, bad and good, with two billion humans.

---

[110] http://en2.wikipedia.org/wiki/World_War_II

Reducing our numbers is not the only thing that has to be done to save what is left.

In his book, *Sparing Nature-The conflict Between Human Population Growth and Earth's Biodiversity*, Jeffery McKee inadvertently summed up the whole overpopulation issue with respect to biodiversity in the first chapter. Having found a termite in his yard while doing an informal animal survey in support of writing his book, he immediately called the exterminator. The population debate is untenable. Let us give humanity something it wants—the TIFIC—and the rest will fall in place. This contraceptive could be the enzyme that starts a chain reaction, creating a world where people are free to lead lives of their choosing without driving to extinction the other creatures that we share our planet with.

CHAPTER 3

Busting myth busting

A book first written in 1986, and updated in 1998, *World Hunger: 12 Myths* by Frances Moore Lappe, Joseph Collins, and Peter Rosset with Luis Esparza, discusses in detail what the authors believe to be twelve myths of hunger. One of the book's authors, Peter Rosset, is also co-director of Food First. I visited their website and found that this is a power-to-the people organization, dealing with issues like the WTO, land reforms, government corruption, genetically modified foods, organic farming, and corporate greed with a focus on hunger and no apparent emphasis on family planning as part of the solution.[111] I drew this last conclusion having been unsuccessful at finding any mention of family planning on the site.

I can't imagine what it would be like to write a book with three other people. It would seem that getting consensus on issues of this complexity would be nearly impossible. Maybe that is why chapter 3 titled, "Myth 3—Too Many Mouths to Feed," is so full of contradictory statements:

> "Surveying the globe, we in fact can find no direct correlation between population density and hunger."

> "...in many countries much higher population densities would make more difficult the tasks of social and economic restructuring necessary to eliminate hunger."

> "No one should discount the consequences of high population density, including the difficulties it can add to the already great challenge of development."

---

[111] http://www.foodfirst.org/pubs/backgrdrs/1998/s98v5n3.html.

> "...it is eliminating hunger that we are after..."
>
> "It is a myth that "We must slow population growth before we can hope to alleviate hunger."
>
> "We believe that precisely because population growth is such a critical problem, we cannot waste time with approaches that do not work."
>
> "Family planning cannot by itself reduce population growth, though it can speed a decline."

Ironically, at the beginning of this chapter they state, "The question of population is so vital that we can't afford to be the least bit fuzzy in our thinking." This chapter has the feel of having been written by someone with multiple personalities.

After having read the chapter more than once I was still not sure what the authors' position was on family planning.

I sent an E-mail to the Food First website asking for clarification. The response said:

> "...poverty and hunger will not disappear simply through the mass sterilization of women and men in the third world." However, they also felt that "...any poverty eradication program would make family planning freely available..."

Fair enough I say. Here is what I think they are trying to tell us in this chapter:

1) They acknowledged that a planet with too many people on it is not good for biodiversity: "We do not take lightly the prospect of human numbers so dominating the planet that other forms of life are squeezed out..."
2) Apparently, they *do* want to see population growth reduced. Why they think it needs to be reduced was never made very clear. Maybe when they state that

population growth is *not* the root cause of hunger, they are grudgingly admitting that it *does* exacerbate it. Somehow, this defies common sense when you realize that it would be impossible to have 200 million or more *hungry* people in India if there were only 2 million people living there. Have I missed something?

3) They want to make sure that we do not try to minimize population growth using the wrong methods—mass sterilization of the poor for example.
4) The proper way to reduce population growth and hunger—since the authors believe that both are caused by the same thing—is through "Improving living standards and lessening inequality, including providing education for women..." According to the authors, birth control programs alone—the simple delivery of contraceptive technology—accounts for only 15 to 20 percent of the fertility declines seen.

Essentially, the argument being made is that fertility rates will drop rapidly and voluntarily on their own if the poor are educated and brought out of poverty, so stop with the sterilization and coercive contraceptive agendas, they just cause suffering and hardship. It is quite true that without poverty, a person can always buy food even if war, or drought, or pestilence causes crops to fail. You can have poverty without hunger, but you cannot have hunger without poverty. This logically leads to the answer to the problem of hunger; eliminate poverty.

If that is all they said in this chapter, I would be done, but they said much more and consequently have drawn my critique.

First of all, I am amazed that anyone should have to survey the globe looking for a direct correlation between population density and hunger. Without moving from my chair I can say unequivocally that there is no direct correlation. In other words, you will not always find hunger wherever you find a high population density. New York City, having only a handful of pea patches that could qualify as cropland, has a population density—number of people per unit area of cropland—that approaches infinity. The Netherlands has a very high population density and also has no hunger. Ethiopia has a very low population density yet it has a serious hunger problem. Chapter 3 spends a lot of time telling us that density is not a problem (when it isn't telling us it is) as if it were not already extremely obvious. Density—taken by itself—has nothing to do with hunger.

Also keep in mind that the density of a country is an average. The densities are high in cities and go to zero in deserts and mountain ranges. Using a city like New York instead of a country to illustrate this lack of correlation is perfectly valid because a city, like a country, is simply an arbitrarily assigned boundary that contains people inside. New Yorkers import all they need just like many countries do. Also keep in mind that when famine strikes a country, it is usually only striking a small area contained inside the borders of that country, not everywhere inside those borders. Because of unequal distribution of wealth, there is some hunger inside New York City, as there is in many countries around the world. Equal distribution of wealth is not possible, human nature being what it is.

Overall, the authors have gone way too far in downplaying the contributions of family planning.

> "Those who cling to family planning programs as the answer to population growth might do well to heed the current experience of China."

They go on to claim that the huge reductions in fertility in China were primarily from social reforms. In reality, a major part of China's five year plan in 1970 included contraceptive and abortion services that were extended into the rural areas, as well as the promotion of later marriage, longer intervals between births, and smaller families. This was the first time that targets for population growth rates had been incorporated into a five-year plan. Ten years later—1979—the Chinese implemented their controversial one-child policy. The authors of the *Twelve Myths* point out that the drop in fertility rates since 1979 have been much less dramatic. They use this as evidence that these stringent population control methods were ineffective. In reality, the fertility rates had already fallen dramatically and could hardly fall any further by 1979. The drop in China's fertility rate from 5.6 in the 1950s and 1960s to 2.1 today is the fastest and most drastic fertility decline in the history of humankind.[112]

Few people know much about the one-child policy. Most have a warped view as presented by the pro-life movement or our profit-driven sensationalist media. First off, the name is misleading. It allows for one child per urban family, two children per rural family, and three for ethnic minorities with small populations.[113] You never read that 80 percent of China's children under age 14 have brothers and/or sisters.[114] The Chinese government had hoped that about 30 percent of couples might agree to forgo a second child if the policy called for an ideal family size of one. Enforcement was left up to each of the many provinces and was anything but consistent. This led to the excesses of the sterilization campaign of 1983 when up to 20 million people were sterilized. Of course, in China that translates to only 1.5 percent of their current population. Out of that number, about half were probably voluntary sterilized, leaving 0.75 percent sterilized coercively after having their quota of children.

In 2002, China finally created a new state law that was designed to eliminate all of the abuses that were going on at the local family planning level called the "social compensation fee." Once again, each province was left on its own to pick the size of the penalty. In some provinces it is a payment of 10 percent of the family income to have an extra child, similar to the luxury taxes seen in Europe. In other provinces, the penalty was ruinous leaving couples with no option but to avoid an extra pregnancy or to abort an existing one. Guess which one you read about in the media?

You only hear about the excesses because they make worthy news. You also hear about the Chinese preference for male children and a sex ratio that shows a shortage of women. The one-child policy gets blamed for this when in reality the Chinese have been using infanticide since the third millennia BC to favor boys over girls.[115] The practice of infanticide continued into the

---

[112] http://www.usembassy-china.org.cn/sandt/fertl21.htm

[113] http://www.atimes.com/china/AJ14Ad01.html

[114]http://www.randolphschool.net/academics/departments/history/independent/Asia/China_PD_CUL.html

[115] http://ieas.berkeley.edu/shorenstein/1998.05.html

early twentieth century and has merely been replaced and somewhat exacerbated by the *illegal* combination of ultrasound and abortion. When infanticide was made illegal, some families got bigger simply because they kept having babies until they finally had a boy. The one-child policy is not the problem; the devaluation of women in Chinese society is the problem. The same thing continues to happen in India as well.

China's success in reducing hunger inside its borders by 60 million in the 1990s was in stark contrast to India, Indonesia, Nigeria, Pakistan and Sudan where the number of hungry increased by 60 million. This success was through the combination of social reform *and* family planning. Is it wise to apply our cultural standards to a country that is the same size as the United States, but has almost five times more people and 25 percent less arable land? Keep in mind that this is China, 30 million starved to death there in 1960 because of bungled government policy and they are still adding about 12 million people to their population every year.

The point the authors want to make here is that natalist policies by governments have little effect on fertility rates. People will have as many babies as they think they need. Simply get people out of poverty and they will choose to have fewer children. In general, this has been the latest trend. However, there are several problems with this logic. People don't know how many children they need, most pregnancies are accidents, and there are examples of successful natalist policies out there that refute their claim, Iran being the most obvious. Iran's reduction in fertility rates was done entirely through non-coercive government sponsored family planning and education of both men and women with very little change in income levels.[116]

The authors of The Twelve Myths are convinced that the poor are being coerced and forced and brow beaten to have fewer children against their own best interests and that this is actually causing more poverty and hunger. If that were true, then we would see a spike in hunger for the rural farm families in China but quite the opposite has happened. They contend that family planning without striving for social justice often results in fewer poor people, not poverty reduction. This is true, but what is

---

[116] http://www.mnforsustain.org/iran_model_of_reducing_fertility.htm

wrong with having fewer poor people? There certainly are fewer wealthy people as a result of family planning.

Having said all of this, I may have given the impression that I am in favor of China's human rights abuses. I am not. However, overall, the government of China has served its people well. The abuses that occur are imperfections that need to be hammered out over time. They have tarnished the Government's success and were not necessary for that success.

I wish Food First the best of luck in their desire to end hunger by making the world a more harmonious, fair, and egalitarian place but their tendency to downplay family planning is where they really go astray. By saying that you can reduce population growth by increasing the quality of life, they have inadvertently suggested the inverse; a decrease in population growth results in an increase in the quality of life. Actually, neither statement is entirely true; one begets the other. Nevertheless, I don't appear to be alone in my confusion. I offer as evidence the website for the Rehydration Project.[117] They have renumbered myth 3 to myth 2, and slightly modified its title to say, "There are too many people to feed." They then reword the myth as follows:

> "Contrary to popular belief, overpopulation is not the cause of hunger. It's usually the other way around: hunger is one of the real causes of overpopulation."

In other words, if you just feed people adequately they will stop having children—a complex issue oversimplified to the point of being nonsensical.

According to the American Heritage dictionary, Thomas Malthus was a "British economist who wrote *An Essay on the Principle of Population* (1798), arguing that population tends to increase faster than food supply, with inevitably disastrous results, unless the increase in population is checked by moral restraints or by war, famine, and disease."

The world population has not crashed as a result of population growth outstripping food supply. Exponential technology growth has caused our food supply to increase, keeping pace or even outstripping population growth.

---

[117] http://www.rehydrate.org/facts/hunger.htm

Was Malthus wrong? Well, so far, if you are talking only about mass starvation. You can't definitively say that he was wrong on this count unless you can guarantee that the world will not one day finally outstrip its food supply. Malthus was a deeply religious, elitist, economist. Clearly, he did not envision the technological explosion and free markets that currently feed us and provide us with contraception. But consider this, there were less than one billion people on the planet in his time. Today, about 2.8 billion people, almost half the world's population, live on less than $2 a day and 1.2 billion of them live on less than $1 a day.[118] So, you tell me. Is the world overpopulated? I guess the answer to that question would depend on your definition of overpopulation. To say that someone lives on less than $2 a day is an attempt to describe a level of poverty that most Americans would find difficult, if not impossible, to comprehend.[119] It is hard to believe, but you and I can live on $2 a day. If you don't eat for three days, you will have accumulated $6. That is enough to buy a twenty-pound bag of rice. Now if you can get someone who has a pot and a fire to help you cook it, you will get your caloric requirements met for several weeks. You can then invest in a blue tarp, a pot, maybe a propane cook stove. Find yourself a nice cardboard box and you're on your way—just like 2.8×10^9 other people.

What is more important, the percentages of people who go hungry, or the total number? If you have only 100 people and 13 percent are hungry, you have 13 hungry people. If you have 6.35 billion people and 13 percent are hungry, you have 825,600,000 hungry people. Coincidentally, that was how many people were on the planet when Malthus wrote his essay, and it is also the number of starving (chronically malnourished) people on the planet today.

Is overpopulation a myth? A growing number of people seem to think so, including some of our leading intellectuals. Is there such a thing as a false meme? Are we witnessing the propagation of one? While discussing the combinatorial powers of the human mind, Steven Pinker, in his book *The Blank Slate: The Modern Denial of Human Nature*, briefly discusses why

---

[118] http://www.worldbank.org/poverty/wdrpoverty/index.htm

[119] http://www.infoplease.com/ipa/A0908762.html

Malthus was wrong and mentions Ehrlich's failed predictions. He then goes on to note that findings by the Nobel Prize winning economist Amartya Sen show that famine is almost always caused by political or social upheavals or short-lived conditions (like drought). Welcome to the modern denial of overpopulation.

Why then, do you not always find famine whenever you find social upheaval and drought? Famine is actually rare, but war and drought are not. The upheaval has to be big enough, or the drought deep enough, or the numbers of people needing to be fed large enough, before you actually get famine. That third variable—the number of people involved—is an important part of the equation. When you are talking about famine relief, it is far easier to feed 2000 people than it is to feed a thousand times that many—2 million. It is also easier to feed 2 million people than it is to feed a thousand times that many—2 billion.

Another important variable is time. It only takes a few weeks to starve to death. Any interruption in food supply that lasts more that a few weeks is a disaster. However, malnutrition that leads to reduced disease resistance is the cause of most hunger-related deaths, not actual starvation.

We have all learned to associate overpopulation with famine. This may come as a shock, but like population density, famine by itself has little to do with overpopulation. People who have made that connection usually draw the wrong conclusion, which is that overpopulation is a myth. The problem is the fuzzy definition people have in their heads of overpopulation. Overpopulation is a concept that embodies poverty, malnourishment, and environmental degradation, and when the right factors all come together, one of the results can be famine. Also, population density can be the final straw that tips a population into famine, but without the other variables—a failed crop for example—it cannot stand by itself as the sole cause.

Let me try an analogy. It is assumed that overpopulated roadways cause car crash fatalities. However, upon closer examination, it is found that fatalities are actually caused by human error followed by rapid decelerations from high velocities. The answer is therefore to find a way to eliminate human errors, high velocities or rapid decelerations. Consequently, we have lowered the speed limit and installed air bags. Of course, making our freeways less crowded by restricting how many cars can be on them while holding the other variables constant would also have reduced fatalities. Overpopulated freeways are analogous to an overpopulated world. The effects of roadwork, human error, high velocities, and analogously, droughts, human nature, and political instability are magnified by the numbers involved.

The archaeological record shows that human history has been one long cycle of *local* population growth, *local* environmental degradation, famine, warfare and collapse.[120] Overpopulation has been an inevitable part of human societies since their inception. Warfare and famine have been a big part of reducing our population growth for a long time. Overpopulation followed by societal collapse has been the norm throughout most of human history. One might argue that it was warfare causing the famine and collapse and not the other way around. There is a logical trap awaiting anyone who makes that argument—logic Pinker can appreciate. What could cause evolutionary selective pressure to select for a human nature of aggressiveness that often leads to warfare other than competition for scarce resources? The chronic warfare that we see to this day is evidence of this aspect of human nature asserting itself even in a world with adequate food supplies.

Julian Simon had it backwards. Population growth does not beget technology growth; it is the other way around. The teeming masses of poor people in Africa are not creating new technology; new technology is creating them. Malthus didn't recognize it but food production by his time was already growing increasingly exponential. The ever-improving technologies involved with agriculture and domestication were growing people right before his eyes. Humanity can be likened to a crop that has been richly fertilized by technology. Antibiotics are the analogy for insecticides.

Richard Dawkins proposed that selfish entities called genes are using people to propagate themselves. You could say that technology is using people to propagate itself. Technology also begets more technology. That is what gives it its geometric quality. That is also why it is not very dependent on human population size. It uses human minds, hands, wants, and needs to

---

[120] Constant Battles by Steven A. LeBlanc with Katherine E Register

bring itself into existence. We may someday heed the words of the Terminator, "It became self aware..." The technology available in 1945 was quite impressive. It is likely that the pace of technological growth would have continued almost unabated had world population stopped growing at that time. Economies might have grown as the result of poverty reduction instead of population growth. Our exploding population growth is a side effect of technology propagating itself. The combinatorial power of human minds is very real but a very small percentage of the minds on this planet are participating in that exercise. You cannot seriously argue that those 2.8 billion people struggling to stay alive on less than $2 a day are active participants.

In a historical sense, Malthus was not wrong. That pattern of fighting over scarce resources has only recently been disrupted or possibly put on hold by a technological spurt; or has it? The fighting over oil may have already begun. If our technology does not find an alternative source of energy to replace fossil fuels before we run out, the ramifications will be horrific. Population activists will get to thumb their noses at the economists and Malthus will be vindicated. Lets all hope that never happens. Most certainly, our technology spurt has not proven to be the answer for biodiversity loss and extinction—at least not yet. That is where the TIFIC would come into play, but not in a climate where the reality of overpopulation is being denied by the likes of Pinker.

It would be an understatement to suggest that Pinker is anything less than brilliant—he also has weird hair. At some level, he must suspect that something is up. Early on in his book, he says:

> "Geneticists call us a "small" species, which sounds like a bad joke given that *we have infested the planet like roaches* (my italics)."

He qualifies his comments by saying:

> "The state of our planet is a vital concern, and we need the clearest possible understanding of where the problems lie so as not to misdirect our efforts. The repeated

failure of simple Malthusian thinking shows that it cannot be the best way to analyze environmental challenges."

I agree that simple Malthusian thinking does not have much value with present levels of technological growth. However, hunger and poverty are not what I would call environmental challenges; they are human conditions. Species extinction and habitat losses are what I call environmental challenges and they *are* being destroyed by simple Malthusian principles. In most instances, the rate of environmental destruction has been accelerated by technology.

The three dimensional world is complicated enough. When you throw in the fourth dimension, most people give up. But, you can't ignore time constraints. Population growth is a function of time. Rapid rates of growth outstrip social services and food supplies. It takes time to deplete resources like water, soil, and fisheries. Extinction rates, finding a replacement for fossil fuels, these things are all a function of time. How long can a planet hold up with nine billion people on it? Time is integral to the concept of carrying capacity. Timing is everything. The planet is in a race against time. Fertility rates are falling, our population is growing, and all the while, our biodiversity is disappearing.

Some religionists and pro-life factions are propagating the idea that overpopulation is a myth to promote their views of contraception and family planning, claiming that quality of life actually improves with an increasing population, implying that one way to end hunger and poverty is to further increase our population.

Before long, they will be claiming that the theory of evolution is a myth. Oh, wait a minute, they are doing that. The Discovery Institute—located right here in Seattle—spearheads the latest attempts to teach creationism in our schools.[121] To escape the negative connotations associated with creationism, they champion a new phrase, intelligent design. Breaking with the concept that faith is a keystone to a belief in God, they attempt to use science to prove the existence of God. Their evidence rests

---

121 http://www.discovery.org/

on a concept called reductionism—irreducibly complex systems. What does that mean you should be asking? Just trust me when I say that it is very complex and that very few people who support intelligent design have any idea what it means either. Essentially, it is an attempt to say that an organism's complexity is evidence for the existence of God. At the cellular level, some mechanisms appear to be too simple to have evolved from a yet simpler mechanism. Since the concept of irreducible complexity is too complex for most lay persons, we are all forced to choose our champions and have faith that they understand it. The champion for the atheists would probably be Richard Dawkins, and the one for the creationists, Michael J. Behe. Here are two facts for you, evolution is not a myth, and neither is overpopulation.[122]

The religionists propagate the idea that overpopulation is a myth because of their belief that people are the children of God, and the rest of the planet was put here for our consumption. The Pro-life people propagate the myth of the myth, because they are against abortion and since abortion is the main reason we do not have eight or ten billion people already, being pro-life is synonymous with being pro-population growth. Some economists propagate it because it fits their belief system that growth is good, including economic growth that results from population growth, which creates both labor and consumers. The fact that population growth cannot go on forever is a topic that most economists are loath to address.

How comfortable would the world class intellectual Steven Pinker be if he were accused of sharing the same viewpoint about overpopulation with his polar opposite, Rush Limbaugh? Limbaugh often uses the fact that all the people on Earth could fit into the state of Texas as evidence that the world is not overpopulated.

John Stossel has a segment on the ABC show 20/20 where he is supposed to bust myths. A viewer from Texas wrote asking him to bust the myth of overpopulation, that "the world is getting too crowded."

---

[122] Oddly enough, the Discovery Institute has received funding from the Gates Foundation to study transportation issues: http://www.gatesfoundation.org/Pacific Northwest /Grants/Grant-29490.htm

I wrote a long-winded E-mail to ABC where I quoted many of Stossel's comments from this segment and rebutted them as follows—the quoted text belongs to Stossel and you will note some repetition:

"... there's no space problem. Our planet is huge."

Stossel immediately confuses the word overcrowded with overpopulated. The number of people per square mile (population density) has almost nothing to do with the concept of overpopulation. They mean completely different things, like overcooked, and overdrawn. The planet is anything but overcrowded. Most of the surface is covered with water and deserts and mountain ranges for God's sake. He then goes on to tell us Rush Limbaugh's favorite example which he thinks disproves the liberal lie called overpopulation:

> "... in fact, we could take the entire world population and move everyone to the state of Texas, and the population density there would still be less than that of New York City."

This Limbaugh factoid works as designed. It gives most people the *feeling* that 6.35 billion people isn't so many after all. I am laughing as I write this. It is hard for me to believe that Limbaugh and Stossel both think that overpopulation is the amount of space available to each human being. If you have ever driven the length of Texas, you understand the meaning of empty. What this example really should drive home is how very little of our planet can support human life. Most of Texas is empty because it would be difficult or impossible to live there. There is no water and no way to grow crops. You cannot easily live in the middle of the Texas deserts any more than you can easily live on the moon. There is a lot of uninhabitable space on this planet and in this solar system. Similarly, a drive across the Midwest would show you corn and soy fields a far as the eye can see with nary a human in sight. To understand overpopulation, you have to see before and after satellite photos to visualize what kind of damage 6.35 billion humans can do to a planet.

Here are some other ways to look at it. How big is a billion? Working non-stop, twenty-four hours a day, it would take me over thirty years to count to one billion. Christ was born a billion minutes ago. The universe is 13 billion years old. Here is another one I just cranked out on my calculator. Picture 6.35 billion people standing shoulder to shoulder along a road. You start driving 60 MPH down that road. After driving non-stop for two full days you can't see the end of the line. After a month, you still can't see it. A year goes by, then two years, still you have not come to the end of the line and you won't for another 2.5 years. It would take you 4.5 years driving 24 hours a day at 60 miles per hour to get to the end of that line. One more, it would take me 190 years just to count those people.

> "... starvation is caused by things like civil war and government corruption that interfere with the distribution of food."

Planetary redistribution of resources, like food and water, will never be close to100 percent effective. Distribution costs, competition, war and corruption are facts of life in all human cultures, like drought and disease. When they show up, their effects are exacerbated by overpopulation. War and drought do not always cause famine. The chances that they will cause famine go up with a high population burden—overpopulation.

> "The good news is with more people, we also have more smart ideas."

We also have more people poaching mountain gorillas for bush meat, rhinos for their horns, elephants for their tusks, sturgeons for their eggs, bears and tigers for their internal organs. Stossel walks into his own trap when he says this because it proposes that the answers to the world's problems is a bigger population—forty billion people being inherently better than 6.35 billion.

Most new technology comes from the developed world, where people have lives that allow them to think, be educated and create technology. The other four billion people exist because of that technology; the technology does not exist because of them. Obviously, you only need a billion or two people to make technology like atomic weapons, antibiotics, and vaccines because we had them in 1945. The exponential technology growth of the 40's would have continued almost unabated if our population had stopped growing then. In fact, our world economy may have grown as a result of an increased standard of living in the world instead of as a result of a growing and desperately poor population.

> "Every year we learn how to grow more food on less land. Thanks to improved technology, the United Nations now says the world overproduces food."

It is time to stop stating the obvious. The problem is, and always will be, that the food is not where the hungry people are. The hungry cannot afford the food and farmers can't afford to grow food and then give it away. That is the reality of the world. Somebody has to pay to grow and move the food. As harsh as this sounds, it is not part of human nature to sacrifice our homes and the wants of our own children to feed strangers on the other side of the planet. Look around you. Do you see anyone doing that? Can you name one person who does that? Do you think Stossel does that?

One can argue until hell freezes over as to why three billion of us live in abject poverty. You cannot, however, argue that the wants, desires, and needs of billions of people are pushing other species to extinction. Our world is being wiped clean of its biodiversity or maybe that is a myth also.

I don't watch much television. I had never heard of John Stossel so I clicked on his book advertisement on the ABC website. His book is called *Give Me a Break-How I Exposed Hucksters, Cheats, and Scam Artists and Became the Scourge of the Liberal Media*. It seems strange to me how a television celebrity is qualified to bust myths of any kind. Will we soon have to listen to Stossel bust the myths of evolution and extinction? Will he one day patiently explain to us that dinosaur bones are the result of Noah's flood?[123]

Back in 1994, yet another conservative—this time the popular political humorist P. J. O'Rourke—wrote a book with the funny title *All the Trouble in the World: The Lighter Side of Overpopulation, Famine, Ecological Disaster, Ethnic Hatred, Plague, and Poverty*. One of his most popular observations was his comparison of the living conditions in Bangladesh with those in Fremont, California. Once again, because both areas have the same population density he concludes overpopulation to be a myth. Starting to get old? Yes, except at least this time it is a comedian who has confused population density with overpopulation.

This repetitious confusion of the two terms suggests to me that the issue is just too complex for most people to grasp, not that they couldn't grasp it if they were well informed and made an effort. Theoretically, there are two ways one could end the poverty in Bangladesh. You could somehow create a thriving free market industrial economy that would make the populace wealthy enough to import goods, or you could lower birth rates and over time reduce the number of people. For the sake of discussion, let's pretend that we flew 209,000 Bangladeshis to Fremont (209,000 is the current population of Fremont). In other words, we just doubled their population. This is an example of a very rapid population growth rate. The citizens of Fremont would experience population pressure and their quality of life would plummet. Inversely, if we were to cut the population in Bangladesh in half, they would experience population relief, and would see an increase in the quality of life. You get two choices to improve the lives of the people in Bangladesh, one would consume a great deal of the planet's resources, and the other would not. Obviously, creating thriving industrialized economies is easier said than done, and so is population reduction.

There will never be a consensus on whether or not human overpopulation is a reality. It is a matter of definition. If you

---

[123] Interestingly, the Discovery Institute (mentioned earlier) just sponsored Stossel here in Seattle to promote his book.

think that overpopulation means overcrowded, as all the aforementioned people do, and since there is no correlation between population density and standard of living, logically, you must conclude that overpopulation is a fabrication. Your logical conclusion, however, is flawed because it mistakes population density for overpopulation. Mass extinction, environmental degradation, overpopulation, and the theory of evolution are not myths being churned out by liberal liars.

The American culture appears to be polarizing more and more into two groups, liberals and conservatives. Once people identify strongly with a group they tend to see other groups as evil—human nature. It seems like every other sentence that is uttered by Rush Limbaugh contains the phrase liberal *lies*. Then there is the book, *Lies and the Lying Liars Who Tell Them: A Fair and Balanced Look at the Right* by Al Franken. Bush is fighting off attacks that he *lied* to the American public about intelligence reports. Excluding profanity, I don't know of a stronger word than liar. Where do we go from here? Do we need to invent an even stronger word than Liar? Isn't it possible that the intelligence reports were wrong and America went to war by accident? Come to think of it, maybe it wouldn't be smart to admit that either. Although, not pertinent here, it is obvious that the Bush administration felt that we needed to go to war and they simply chose to believe the reports that gave them an excuse for doing so. They could have chosen to believe the other reports that suggested that the Iraqis had no weapons of mass destruction.

It is a crazy world. In the end, people will either support contraception and a woman's right to choose, or they will not. World population rests on those issues. Today, the protection of the Earth's biodiversity is largely in the hands of NGOs. Our population will stop growing. How big it will be when it stops and what happens next is unknown. Efforts to protect those areas harboring our biodiversity will eventually fail unless the number of people wanting those resources decreases.

A definition of overpopulation is needed at this point. I don't want to pick a new word for overpopulation. Sometimes, as we humans go about the business of being human—disagreeing, arguing, fighting—one group or another will change their name in a futile attempt to escape a negative connotation that has become associated with that name. Atheists have been trying to lose the word atheist every since Darwin's bulldog, Thomas Henry Huxley coined the word agnostic.[124] An atheist can now pick between agnostic, secular humanist, free thinker, or bright—recently coined by Richard Dawkins. Homosexuals chose the word gay. This upset some people because it robbed the English language of a common and useful word used to describe someone who was fun and happy. ZPG has changed their name to Population Connection, and the Moral Majority changed their name to the Liberty Federation.[125]

There are several definitions of overpopulation floating about out there. Hopefully this exercise will help clarify things should readers have different definitions in their heads. The definition for overpopulation is very straightforward when applied to elephants trapped in the confines of an African game park. Likewise, for people trapped on a small island as happened on Easter Island. It has been very straightforward for most of human history as one society after another—confined by geography or the boundaries of competing societies—exhausted the natural resources within their boundaries.

I want to start with a true story about a place that was once overpopulated. As is typical, environmental degradation was occurring at a fast pace. Water resources were being depleted and polluted and water borne diseases were on the rise. Conflict over those resources was also on the rise. The problem had been exacerbated by a drought. Crossing the border in an attempt to find food and water was asking to be shot by the people of that land.

A think tank was formed in an attempt to come up with a plan. It started with a brainstorming session. One idea was to bring more water in with trucks or a series of canals. Another idea was to supplement diets by bringing in more food. The idea of increasing available food by planting high yield, pest and drought resistant crops was kicked about. Seeding the clouds to make it rain was considered—that was a ridiculous idea. Some-

---

[124]http://www.infidels.org/library/modern/reason/agnosticism/agnostic.html
[125] http://religiousbroadcasting.lib.virginia.edu/pubs/whykill.html

one suggested that a way had to be found to convince the people in the neighboring territories to allow the famine victims to move onto their lands to share their food. That was an even more ridiculous suggestion. Finally, someone suggested the implementation of a newly discovered, long-lived contraceptive.

After mulling their options over, they had to decide which methods to use. "No point using the contraceptive," one of the think tank members said. "Hunger is caused by injustice, not overpopulation. Overpopulation is a myth. We have to find a way to make their neighbors give up some of their food. It is immoral and unjust that they should be allowed to sell their excess crops for profit when starvation is happening right across the border." The priest lost the argument however, because the other tank members thought it would probably be a hell of a lot easier to just use the damn contraceptive than to build canals, grow special crops, control the weather, or, especially, convince people to share their food with interlopers from another land. Besides, what were they going to do ten years from now when the population is even bigger?

This story actually took place, although I embellished it a little with that piece about the priest and the think tank. It was an African game park. The individuals involved were elephants. The border was a fence meant to keep them in. Their neighbors who refused to share their food were the farmers that surrounded the park. The contraceptive vaccine was administered with darts and was effective for about a year. Alas, the plan was eventually abandoned because the cost of vaccinating 3000 elephants every year was too high. They would have to track, mark, and dart ten elephants a day. It was much cheaper to kill them and sell their meat, so they did, and almost everyone lived happily ever after.[126]

What is missing from most definitions for overpopulation is the requirement for a boundary. Overpopulation, like poverty, is a local phenomenon, and, as with poverty, it waxes and wanes with conditions and time. It disappears when a population crashes or if you find a way to feed and house the people. The definition of overpopulation is difficult to apply to an entire

---

[126]http://www.guardian.co.uk/conservation/story/0,13369,1032721,00.html

planet, which has the vacuum of space as a boundary. Overpopulation does not exist within the boundaries of all societies, so defining the entire planet as overpopulated is a kind of average. When relief agencies exist and are free to leap across national boundaries to nip famines in the bud, you have defeated one of the age-old problems with overpopulation. You can keep people from starving until the conditions that tipped the population into famine abate; a war or drought may end, the population may decrease from mass migration as happened in Ireland, a blight resistant potato may be introduced. Likewise, free trade allows societies without enough natural resources within their boundaries to remain fed and housed. Since overpopulation seems to share so many characteristics with poverty, maybe we should start there.

I found the following definition on the internet at a website that promotes the belief that overpopulation is a myth: "… overpopulation should be defined as a situation where it is impossible to provide everyone on Earth a living standard at the subsistence level."[127] I don't like it because it requires famine before you can qualify an area as overpopulated. I find two other flaws. One is the definition of subsistence, which is arbitrary. The other is that, as I mentioned earlier, overpopulation exists inside boundaries and this definition assumes that "…everyone on Earth…" must all be in a simultaneous state of overpopulation. Personally, I prefer a definition that says you have overpopulation whenever you have people *living* at the subsistence level. If you assume that people who live on less than $2 a day are at the subsistence level, then roughly half of the people in the world are living in overpopulated boundaries. Depending on how you define subsistence, you may find three-quarters or more of the boundaries are overpopulated. Using this definition you can safely say that most of the World is overpopulated.

Ehrlich has a definition of course. It recognizes that overpopulation does not happen instantaneously. It takes time to deplete resources. That is why his definition incorporates the concept of sustainability. If you are using up your resources faster than nature can replenish them, you are in a state of over-

---

[127]http://www.overpopulation.com/faq/basic_information/what_is_over population.html

population and it is only a matter of time until famine hits or grinding poverty and poor nutrition lowers life expectancies. Sounds like a description of much of the world as we know it today. Famine is not a necessary condition to have overpopulation. Overpopulation can also be defined through resource depletion, biodiversity loss, and deep poverty.

If you type the words "overpopulation myth" into your search engine you will discover that the Catholic Church and other pro-life factions have latched onto the idea that overpopulation is a myth. When a United Nations report emphasizes that "population is only one of a number of complex, interrelated issues affecting the environment and human development," some religious groups interpret it to mean that overpopulation is a myth.[128] In their eyes, abortion, euthanasia, and overpopulation are all closely linked. In that, they are mistaken. Individual women desperate not to bring another child into the world use abortions. With the exception of China, abortions have not been forced on people as part of an orchestrated effort to reduce population growth. Just as with the TIFIC, the fact that abortions also reduce population growth is an unintended side effect. Euthanasia is a choice taken by terminally ill individuals desperate to avoid the prolonged physical and mental agony that usually accompanies the act of dying in our modern world. It is also not part of a master plan to control population. In addition, euthanasia is too rare to have a measurable impact on population.

Admittedly, *forced* abortion, *forced* sterilization, and *forced* euthanasia have all been proposed as a way of controlling population by extremists in the population debate, but those individuals are few and far between and have done far more to hurt the cause of overpopulation than to help it.

If you want to maximize the attack on world hunger, you go after both concerns simultaneously, allowing those who are more interested in the social inequality side of the equation to do their thing and those interested in the family planning side to do theirs. Attempts by one group to invalidate the other only hamper efforts to improve the quality of life for the poor and make enemies out of those who should be allies.

---

[128] http://www.euthanasia.com/popmyth.html

CHAPTER 4

Population and poverty

Population *growth* is a measure of how many people are added to the world every year. It is a rate. If the rate is too high, then increasing the efficiency of food production and distribution—inventing new food technology and implementing social or land reforms—fast enough to keep people from going hungry is a very real concern because social reforms and the development of new technology take time. That is what we have witnessed in the last three decades. In 1970 over a third of the world was hungry because population growth had outpaced social reform and farming technology in countries that needed them. In theory, hunger should disappear if social reforms and technology catch up with growth. Clearly, a growth rate that is too high is a problem, although a transient one, because growth must eventually stop. However, if the growth isn't stopped soon enough you will end up with a huge number of people and will have to wrestle with issues like sustainable living and a world where the majority of the people live in grinding poverty—which is where we are today.

Population, poverty, and hunger are all interrelated. There is no doubt about that. They are variables that describe quality of life. It is obvious that there would not be a quarter of a billion hungry people in India if there were only a million people living there. Presently, a thousand times that many—over one billion—live there.

The idea that we can eliminate poverty and raise the standard of living for billions of people for a generation or two until their birthrates decline is admirable, but problematic. Poverty reduction is a complicated issue. During the sixties, President Johnson declared war on poverty in the richest country on Earth and made almost no headway. Just exactly how does one eliminate poverty? If you know the answer, please tell the welfare department.

To bring the standard of living up for everyone in the world in the next fifty years to that seen in America today, we would need an increase in the world economy of about four times, have a planet two times larger, or have a few extra planets. The people of the world cannot consume four times as much energy, water, and natural resources. This concept would most certainly finish off what is left of the planet's ecosystems. Logic dictates worldwide population reduction to achieve worldwide poverty reduction without chewing up what is left of the planet's biodiversity. People do not simply stop seeking status and wealth when their net worth hits some innate value. The only thing that stops us from obtaining more is the fact that, eventually, we can't obtain more. We only stop when we are forced to do so by circumstances—our income plateaus, we can't afford to go deeper into debt, or our health fails. Once a Chinese peasant can afford a scooter, he will obtain one, and when he can afford an SUV, he will obtain one of those too. The instinctive urge too continuously seek higher status does not satiate itself.

There is a mindset that is the very definition of poverty. The word hopelessness best describes it. When your day-to-day life is hard, when one's future looks bleak and hopeless, the discipline needed to deal with clumsy, or expensive contraception techniques drops away. I can say this from personal observation, having been surrounded by poverty as a child and adolescent. You cannot say that poor women in the United States have more children in order to staff the farm or to increase the odds that some children will be around to help in old age. Our culture does not work that way anymore. The TIFIC would be an equalizer between poor women and wealthy women. Without the burdens created by multiple unintended pregnancies—time, energy, and money—poor women stand a vastly better chance of breaking the poverty cycle, which I know from personal experience can be very difficult to do. They would be better prepared to compete for their share of the pie, thus forcing a redistribution of wealth.

The argument over whether reducing poverty causes a decrease in fertility rates or reducing fertility rates reduces poverty reminds me of two people arguing over whether a glass is half full or half empty. Raising the standard of living for billions of people is much easier said than done, especially if you try to do it without family planning. Certainly, where raising the standard can be done, it should. Adequate family planning is often the key that unlocks the chains holding people in poverty thus allowing them to find their own way out rather than having a government bureaucracy try to lift them out.

The TIFIC would dissociate poverty from high unintended pregnancy rates. The TIFIC would end resistance to family planning just because abortion is presently part of it. The TIFIC would allow the business of poverty reduction to proceed uninhibited by family planning debates. The Vatican will still argue that we should not be having sex for the fun of it. Those who want to follow that doctrine are as always free to do so. However, when temptation gets the best of them, a protein in their blood that will protect them from pregnancy will be a good thing.

Available natural resources vary from location to location around the world. For the sake of discussion, let us imagine that rapidly changing weather patterns cause India to become as dry and parched as the Sahara. You would expect to see an increase in the number of hungry. Clearly, population size in relation to available natural resources is a concern in countries that are dependent on their own agricultural base to feed themselves. It is easy to see how a country that has something to trade, like oil or manufactured products, can thrive in a desert by trading for all of its needs. You can say that they are not overpopulated although they may be very densely populated. However, not every country has manufactured products or oil to trade. They rely on their close-at-hand natural resources. Decrease those resources or increase the number of people depending on them and you have overpopulation. This concept is not rocket science.

As a side note, my dictionary printed in 1970 does not contain the word overpopulation. Will it also be missing from dictionaries when my youngest daughter goes to college nine years from now? We can only hope.

I have one last example. Picture a mother with one-child who can barely find enough to eat for herself and her infant. One day she realizes that she is pregnant again. Her small world just became overpopulated. That brutal scenario plays itself out every second of every day—literally. Now imagine this impoverished woman finding herself in this situation six or more times in her lifetime. Population is a factor in poverty and hunger. No one will deny that overpopulation scenarios exist in the rest of the animal kingdom. So why are some people trying to say that overpopulation is a myth for Homo sapiens? That anthropomorphic point of view looks familiar to me. The Christian religion has the belief that humankind is not part of the animal kingdom; we belong to the kingdom of God. It makes me wonder. Part of the problem may lie in the propensity to confuse the definition of overpopulation with population density, but I feel there is much more to it than that. Any idea claiming that the quality of life will get better with a bigger population holds much appeal to those who oppose abortion or contraception.

Overpopulation is a dynamic concept. It is in a constant state of change, growing in magnitude inside a geographic boundary because of certain variables at one time (more people, drought, war) and becoming less of a problem due to changes in those variables later (fewer people in a given boundary, or more resources to share in that boundary). Although world population is increasing, the percentage of hungry people has dropped. This statistic is considered proof that population size does *not* exacerbate world hunger. Inversely, this line of reasoning implies to some that the answer to world hunger is to increase our population. This is revisionist thinking—taking a fact and twisting it to suit one's ideology. The percentage of hungry people has dropped because of the green revolution, intense efforts by the UN and NGO relief organizations and mostly because of China's desperate but successful policy to slow population growth. An analogy: Traffic accidents have increased in the last twenty years, yet fatalities have gone down. Obviously, to further reduce fatalities, we need to increase the number of accidents.

The Food and Agriculture Organization of the United Nations (FAO) was formed in the 1940's. The first world food conference was held in 1974.[129] At that time, there were more hungry people on the planet than at any time in history. Intense

---

[129] http://www.fao.org/UNFAO/histo-e.htm

efforts to reduce fertility rates and to improve the plight of the hungry started gaining momentum. The World Food Summit was held in 1996 to review the progress that had been made over the past twenty or so years. Things were looking much better. There was 15 percent more food per person than for the 4 billion people of 20 years ago and the number of hungry had been reduced from 950 million to 850 million.[130]

Looking at these numbers from a realistic perspective, you will find that they are not very comforting. For example, after more than a quarter century of intense all-out effort to reduce hunger in the world, we have only decreased the number of hungry people by about 100 million. A war, a global economic downturn, or a prolonged drought could turn a number that small around in a matter of months.

The fact that there is 15 percent more food per person in the world is tarnished by the fact that Americans and many others in the world are 15 percent *heavier* than they were in 1974. The number of overweight people now equals the number of underweight people.

In 2002 the FAO stated, "... progress in reducing hunger in the developing world has slowed to a crawl and in most regions the number of undernourished people is actually growing."[131] It makes me wonder. If our population had not grown since 1974, would hunger have been eliminated from the planet a decade ago?

When you type "causes of world hunger" into the Google search engine, you will find a plethora of websites that deal with the issue of hunger. Inside these sites you will find terms like "too many people," "population size, and my favorite, "population-resource imbalances" but you have to look long and hard to find the actual word "overpopulation." I think it's fascinating that a word can be shunned so universally so quickly. Overpopulation has a great deal to do with environmental degradation, soil depletion, deforestation, fishery depletion, and water borne diseases. Are we to assume that these things do not exacerbate poverty and hunger?

---

[130] http://www.news24.com/News24/World/News/0,,2-10-1462_1450334,00.html

[131] http://www.fao.org/DOCREP/005/Y7352e/y7352e03.htm#P34_5028

It is obvious to me that the word overpopulation has become politically incorrect in the politics of hunger. Political correctness stifles intellectual discourse. Discussions of the Copernicus system of the planets became very politically incorrect in Galileo's lifetime but eventually held sway.

The fact that the word has disappeared is indicative of something very important. I am not sure what, but I do have my suspicions. A small number of overpopulation extremists have proposed draconian measures to control our population by allowing the death rate to increase to what they call a natural level. How would they do that you might ask? Well, the ideas usually involve things like forced abortions, euthanasia, eugenics, and worse.

Family planning advocates hate to see these guys show up at their Internet forums, as do proponents of euthanasia as practiced in Oregon. Population extremists are the negative stereotype most often associated with the word overpopulation. Their less than popular points of view follow them around like the grim reaper, casting a shadow on groups who are just trying to help the terminally ill or insure the reproductive rights of women.

As a matter of fact, Ed Glaze, one of the editors for this book, is one of these population extremists. He is motivated by the "concern that socially acceptable methods of population control will not be sufficient to avoid catastrophic collapses of the economy, society, and the environment." He posted on his forum site excerpts from a book by Charles C. Wise called *Beyond Love: Will Mankind Be Tough Enough to Survive the 21st Century?* Glaze highlighted the following sentence for me:

> "What are a few billion premature deaths of redundant human beings when weighed against saving this beautiful planet and its many wonderful varied life forms?"

Here is another one I found that I thought had sufficient shock value to be mentioned as well:

> "If our world is to be saved, hopefully the needed reductions in numbers (at least in the industrial nations) can be achieved without the deliberate destruction of whole

classes of people (other than criminal) just to get rid of them."

If this mindset sounds familiar, you are not alone. Glaze created a folder at his forum site called "Comparisons to Hitler" where he collects all such comparisons to himself from people who visit his forum.

Who would be classed as a criminal? Well, for starters, illegal immigrants. People desperate to escape poverty and improve their lives, sneaking across our porous borders—just as I would do if the tables were turned—would be put on the list. You can now see why there has been such a backlash against the movement to reduce population growth. We have a handful of extremists to thank for that. Glaze, of course feels that by expressing his point of view he will convince others of its validity. Instead, his views have the opposite effect; they steel resistance to reducing growth; they pour gas on the fire, not water.

You might be questioning my wisdom at this point. Why would I accept a critique from someone with such potential to stir deep controversy? My reasons are multiple. First, Glaze is an authority on the subject of overpopulation and hosts an organized and well-run Internet forum for discussing population issues.[132] Finally, controversy stirs debate and therefore, thought. A book devoid of controversy makes for dull reading. I want to conclude by stating clearly and unambiguously, that I do not support his solution to our population problems in any way, shape, or form for a multitude of well defended, moral, and highly rational reasons. I would strongly advise others not to put stock in his ideas as well.

Another idea I have seen to control our population is to hold our food supply constant—gradually decrease the supply and then not allow it to grow again. This concept came from a study of monkeys in the wild to determine the mechanisms that control their population. The study found that when the natural food supply was constant—not increasing or decreasing—famine and malnutrition were absent. Their findings seemed to indicate that a slow steady birthrate seemed to keep pace with death rates. An increase in death rates through disease and predation was not

---

[132] http://forums.delphiforums.com/overpopulation/start

detected. Although the mechanism behind this perfect timing of births was also not detected, the researchers suggested that holding the human food supply constant might be a means of controlling our population.[133] I wrote a long critique of this study but chose not to include it in this book. The author's proposal that we might control our population by simply holding the world food supply constant based on these incomplete observations is absurd and naïve. Any student of history would predict that a bureaucracy with the power to control the planetary production and distribution of food would quickly become an Orwellian nightmare.

It is easy to see why overpopulation extremists get the lion's share of the credit for burying the word overpopulation. The one-child policy in China was also universally unpopular, as were some other failed attempts to control birthrates in India.

There is another reason—religion. The Catholic Church has historically resisted contraception and has always found the whole idea of overpopulation repugnant. The pro-life movement is almost exclusively religion based, and they too find the topic of overpopulation distasteful. It is difficult to say why a term or concept becomes unmentionable, but in this case, I suspect that many people associate the concept of overpopulation with being pro-choice. Following in the missionary tradition, most of the movers and shakers in the movement to prevent hunger and poverty are Christian-sponsored. Bread for the World is one such group. The Bread for the World Institute is a nonprofit organization spun off the original group.[134] Neither group has an official position on abortion as far as I can discern. Their focus is strictly on hunger and poverty. All the same, if your boss and supporters do not want to hear the word overpopulation, it will have a tendency not to appear on your reports, and that is exactly what has happened. The idea that overpopulation has nothing to do with hunger or poverty has become an urban legend, like the one that says there are more trees in America today than when the pilgrims first landed.

On many hunger sites, people concerned about the effects of overpopulation are labeled as pessimists, or neo-Malthusians.

---

[133] http://www.oilcrash.com/population.htm

[134] http://www.bread.org/whoweare/index.html#whoweare

Supporters of Julian Simon's view are called optimists.[135] Something is going on here. Where are the realists?[136]

I don't like being called a neo-Malthusian. It falls into the same category as Marxist or communist, or atheist. Besides, few people have any idea what a neo-Malthusian is. They also have no idea what a Marxist or a communist is, for that matter. Such Zealots are evil; that's about all they know for sure.

It is derogatory to call someone a neo-Malthusian just because they support family planning or strive to reduce population growth. In addition, Malthus sounds too much like Malfoy, the evil family in Harry Potter—"mal" meaning abnormal or bad.

In order to isolate myself from the Malthusians, pessimists, and the population extremists, I'm forced to invent a new, more precise word—*poperty*. Yes, it's silly sounding, but I like it because it will be easy to remember and because it takes some of the seriousness out of the discussion. Poperty combines the definitions of overpopulation and poverty.

Poperty: when competition from other people causes the condition of being poor—a lack of resources for reasonably comfortable living. Extreme poperty implies that sustenance such as food and shelter is lacking. Poperty can be described with an equation: poperty = A/B, where each unit in the numerator A represents one person and each unit in the denominator B represents the amount of available resources needed by one person to achieve a reasonable average lifespan. Lifespan is generally considered a good indicator for quality of life. The optimal value for the poperty quotient is one. A value greater than one indicates a tendency toward a shortened average lifespan caused by malnourishment, increased susceptibility to disease, or lack of shelter. A value less than one indicates a tendency towards opulence. The value of poperty is bounded on the upper limit by death from starvation, disease, exposure, or a combination thereof. It is bounded at its lower limit as the value approaches zero. The value can never reach zero but, as witnessed by the extreme opulence seen with wealthy individuals, it can get very small indeed.

---

135 http://www.ifpri.org/2020/briefs/number19.htm

136 http://www.worldhunger.org/articles/fall2000/messer1.htm

Poperty is very similar to poverty except that the number of people is an important variable in its definition.

I designed an experiment in my head to check my definition of poperty. A fence encircles a plot of fertile farmland with food crops ready for harvest. Included on this land are a fully equipped barn and farmhouse capable of processing the crops into food. A well with clean water, an outhouse and a pond stocked with food fish are also on the premises, along with a number of trees that can be used for fuel.

One volunteer is added to the enclosure each week until the poperty exceeds a value of one. Researchers will know this point has been reached when malnutrition starts to set in. At this point, you can continue to add people, as happens in many developing countries, or stop and wait to see what happens. What I would expect to see over time is males vying for dominance followed by conflict and violence.

This made me suspect that poperty may have an element of reversibility. Conflict, violence, and warfare can create poperty as well as be created by it. Poperty also has the property of being scalable, making it applicable to small villages and entire countries as well.

The variable in the denominator called B has food as one of its biggest players. You can have adequate housing, clean water, and protection from disease, but without food, the value of B will drop, causing poperty to increase. You can have adequate food, but not access to clean water or sanitation, thus increasing your susceptibility to disease. This too, can cause the value of B to shrink, again causing poperty to increase. You can juggle the values of A and B to get the value at or below one.

When famine strikes an area it is too late to shrink the value of A—the number of people in the area involved in the famine. You can only increase the value of B by providing food and shelter. However, a way to minimize the chance of famine for a given boundary in the future is to decrease the value of A.

How many children would be dying of malnutrition-related causes today had China's fertility rate remained unchanged? Millions of children around the world live short lives in abject poverty because their mothers never had the option of participating in family planning. Ironically, many of the same people trying to send them food also voted for a U.S. president who reinstated the global gag rule that cut off UN funds for family planning.[137]

There are two basic mindsets out there: those who concentrate on minimizing poverty and hunger, and those who concentrate on preserving the planet's ecosystems. One group's goals cannot take precedence over the other. The goals of both groups must be achieved. Let's face reality. It would be a lot easier to reduce poverty and thus hunger, and protect the biodiversity of the planet if we could somehow keep our population from swelling all the way to nine or ten billion.

Both groups are for the elimination of poverty and hunger, and both groups are for the preservation of the planet's ecosystems and biodiversity. This division of labor to improve the plight of the world's inhabitants is efficient and positive except in the cases where one group's efforts nullify those of the other. An example of this conflict of interest is the building of hydroelectric dams. These projects are usually successful at reducing poverty, but they also wreak havoc on river ecosystems. The most notorious of these projects is the Three Gorges Dam on the Yangtze. It will be the largest hydroelectric dam in the world. It may well be the final nail in the coffin for the hundred or so remaining Chinese river dolphins—another extinction on the way. There are concerns that the buried cities behind the dam will turn the reservoir into a giant cesspool.[138] This dam is an example of the price paid by our ecosystem to decrease poverty. Without a shrinking population, it is a trap that we cannot escape.

Some believe that the problem of global hunger is one big paradox. There has been enough food on this planet to adequately feed every human being since the 1970's. In addition, most of the hungry live in rural areas of the developing world and subsist primarily by growing their own food. I am not one who sees it as a paradox. It may be ironic, but that is different from being paradoxical.

People appear to go hungry amid plenty because they really are not amid plenty. The food being alluded to is not *amidst*

---

137 http://www.ippf.org/newsinfo/pressreleases/2001_06_11.htm

138 http://www.cnn.com/EARTH/9711/04/china.dam.reut/

them, it is usually hundreds or thousands of miles away and is the property of someone else. That apparent paradox was easy to solve.

When one accounts for human nature and the real costs of distributing food, it is obvious why excess food in one part of the world does not get to the parts of the world that need it. All of us in America who want to end world hunger, live lives of extreme opulence compared to most people on the planet. Given the choice of sending our children to college or taking that money and donating it to food relief, we send our children to college—every one of us. It is instinctive; it is called human nature. It is very real and has to be accounted for. We cannot help but to care for our loved ones first. Caring for strangers on the other side of the planet is very low on our genetically programmed list of priorities. It is that way for all human beings. We will never find the answers to ending poverty and extinction if we continue to deny that there is a thing called human nature.

It is easy to find numerous examples of countries that export grain while their citizens are going hungry. At first thought this sounds barbaric. However, if you keep thinking, you can see why it can happen. Most governments don't own their agricultural industries. They usually regulate them to some extent, but farm products are part of the free market; that food belongs to the farmer who grew it, and he is free to do with it what he wishes. Communist governments are about the only exception. Individuals grow grain to feed themselves and to make a profit with the excess. It is the same reason we all go to work everyday. If that grain is exported, it is because someone bought it. Just for the sake of irony, let's say it was bought by a food relief agency to give to famine victims in another country instead of by cattle farmers planning to convert it into another kind of food. It is still going into the human food chain one way or another.

Now, this impoverished government could have stepped in and forced the farmers to *give* the grain to the hungry in their own country. Those farmers might be bankrupted by such a move because they went into debt to grow those crops. They did not grow them for fun. They must cover their costs. The government would now have to find a way to prop up its bankrupted farms. Had the relief agency not bought that food, someone else would have gone hungry. Another relief agency might at that very moment be buying up food surpluses in America to help this same country with its hunger problem even as its farmers are exporting their grain. Our government could choose to do the same thing. It could force American farmers to donate their food to the hungry people of the world every other year or so. People are people after all. Why should you have to feed the hungry in your own country before you feed the hungry in another?

A government might be making the best move for its citizens by allowing that grain to be exported even though it has hunger inside its borders, especially if it knows that a boat load of free food aid has just docked in its port. This is an example of how food aid can shoot itself in the foot. If the government has the resources to buy grain for its hungry, it might be wise to buy it at a lower price from elsewhere and allow its farmers to export their crops for a higher price instead. Complicated isn't it? Actually, as with any modern economy, it is too complicated to understand fully, or to control. That is why individuals (including stock market analysts) who claim they know all of the answers are to be watched especially carefully.

A universally egalitarian world where all food and wealth is shared equally is not possible. Human genetic programming will not allow it. We are wired to compete. One man's terrorist is another's freedom fighter. One person's justice is another's injustice. To some, justice is giving women the dignity of choosing their family size. To others, abortion is murder and contraception is a deadly sin. Conflict is a part of the definition of humanity. All attempts to create utopia have been disasters. Communes come and communes go. Admittedly, striving to minimize inequality must continue, or we will all end up as slaves. There is plenty of room for improvement, but a balance of power must be maintained or the entire free market system will break down, making matters much worse. The best answer is to continue all efforts simultaneously, especially family planning. Those who think that creating an agrarian utopia is the only answer to reducing hunger are poor students of history. Simply put, struggles to end oppression and poverty would be more successful if there were not so many of us to contend with.

I have heard more than once that we could feed a billion more people if we would stop feeding grain to livestock. This is blatantly not so. First off, there is already plenty of food on the planet. In *theory*, we could feed an extra billion people today without starving our cows. In *reality*, growing extra vegetables in my backyard garden will not help a starving child in Africa, and neither would a mandate that made it illegal to feed grain to cows. The grain simply would not be grown if there were no one willing to buy it. If farmers were forced to grow it anyway, they would also have to be forced to pay to distribute it to all of the places around the globe where people need it. If we discovered that the moon really was made of cheese, malnutrition on Earth would not be affected because the distribution costs associated with retrieving that cheese would be prohibitive. I am well aware of the power of cheese, but not even cheese can fix the world's problems.

What discussion of population would be complete without mentioning Easter Island. Hundreds of years ago the population of Easter Island collapsed in a frenzy of starvation and cannibalism. The island's once pristine ecosystem has been grotesquely degraded.[139] You can view Easter Island as a microcosm of the planet. Not every village on the island ran out of resources at the same time. The shortages started in one village and eventually encompassed the entire island. There were a few people who survived this period, and their descendents continue to eke out a dull impoverished existence there. A similar fate may await the entire planet. That island undoubtedly had its equivalent of modern day Julian Simon supporters. The doom and gloom predictors were also there saying unpopular things like, "we are overpopulating our island."

Surrounded by an ocean, the people of Easter Island eventually consumed themselves into a state of permanent poverty. The Earth is also an island surrounded by the immensity of space. Those NASA nerds who think that the answer lies in leaving this raped planet behind to colonize the stars all have their heads up their helmets. It would be easier to colonize the deep-sea oceanic ridges. It is true that the water is cold down there and the pressure is high, but Mars is hundreds of millions of miles away

---

[139] http://www.apj.co.uk/rapanui/display_article.asp?specifier=history

through a minus 375-degree hard vacuum. This island analogy can be used to make many points. For example, if there was a group lobbying for sustainable lifestyles on Easter island, it apparently failed to make headway before the swelling population overwhelmed its philosophy.

This island's fate was sealed the day it was discovered. There would have been no way to convince those islanders that they should stop having children and competing for higher status, and start living in equilibrium with their limited resources. Without a TIFIC they were doomed to follow their genetic programming which culminated in overpopulation. They were just as we are.

Theoretically speaking, there is plenty of food and water in the universe or in our galaxy or even here on Earth. That has been true for all of human history. When a Neanderthal starved, he did so although vast herds of wooly mammoths were walking about where he was not. There has always been plenty of *potential* food on this planet. The oceans are filled with fish, but until those fish have been caught, processed, and distributed to those who can purchase them, they aren't *food.* Something has to be edible, affordable, and accessible before you can truly call it food. Sunlight isn't food until it is photosynthesized into something edible. Surplus food that is discarded can't really be called food unless it meets all of the criteria. It is only *potential food.* I am going to create a new word here for potential food: *potood.* Silly sounding and hard to forget, I will tie it in with poperty and another word later to create a witty conclusion. If a food is not affordable, or can't be given away, it will fail the criterion for affordability and remain potood. If it isn't where the hungry people are, it fails the criterion for accessibility (which is linked to affordability). Cheese cannot be called food for people who are lactose intolerant. Mammoths in the wrong place at the wrong time were also potood.

One problem with attempting to solve hunger with higher yield pest-resistant crops is the tendency for those crops to increase the amount of potood and the profit margins of those who produce it, instead of feeding those who need it. There is no question that the green revolution has helped feed the world, but if the amount of farmable land is held constant, a slight increase in crop yield for a subsistence farmer is easily overwhelmed by population growth.

Growing and processing one's own food is not usually as efficient as purchasing food grown by free market industrialized farming. I have watched my wife dry cherries and can tomatoes. I ran some numbers to see what we would have to charge for a can of those tomatoes to break even based on what our labor was worth per hour. Purchasing the tomato plants, planting them, watering them, and then canning them took a considerable investment of time. The cost came to a little over 45 dollars per can. Fortunately, she did it for fun, not out of necessity. She could have bought stewed tomatoes for about 49 cents a can. Looking at it from that exaggerated perspective it is a miracle that those poor subsistence farmers ever get enough to eat. Historically, subsistence farming has always been difficult and prone to periodical shortfall. The sheer number of people affected when drought or political instability strikes is the only thing new.

This suggests to me a strategy to alleviate rural poverty and relieve pressure on nature by getting people out of subsistence farming and into cities and jobs, leaving the production of food to larger companies. This is not a new idea, and there are many who think it is a bad idea and it is a bad idea in some cases. There are places in the world where large landowners do not use the land under their control to efficiently produce food. It would be better in these cases to distribute the land to small farmers. In general, however, large farms tied into the free market are much more efficient than small subsistence farms. The advantage of a small family-run farm is that the farmer gets to keep or sell what is produced. It is a way of giving them control, essentially creating self-employment and all of the risk associated with it. A small farm is nothing more than a small business. A farmer is a small business owner. It is a fact of life that there are not enough jobs on the planet to employ every human. You can alleviate that problem with stronger economies or by reducing the number of people needing jobs.

Overpopulation begins as a local problem. Hunger in *Ethiopia* is caused by a shortage of food per person in *Ethiopia*. It has very little to do with the amount of potood in America. You only have two choices to fix the problem. Put more food in *Ethiopia* or fewer people in *Ethiopia*. The math analogy for hunger is very similar to the one for poperty. A hunger quotient equals the number of people in a given area divided by the amount of food in that area. You can reduce famine by increasing the denominator or decreasing the numerator. The area we are talking about does not include the solar system or the entire Earth, just the immediate area where the hungry people are. Keep in mind that decreasing the number of people in America doesn't help and neither does increasing the amount of potood in America. Countries without things to trade for food—goods, oil, money—must grow their own. It is much harder to grow food in sub-Saharan Africa than it is in the American Midwest.

There are only two ways to look at it. You can call hunger the result of overpopulation if you think that the problem is caused by the numerator being too large. If you choose not to look at it from that perspective, then you will say that hunger is not caused by overpopulation, but by the shortage of food (the denominator).

If you agree that human nature prevents us from sharing a significant amount of our wealth with strangers, then we cannot always increase the denominator enough to fix the problem by shipping food surpluses around the world, or by dividing property and wealth up in an egalitarian manner clear across the planet. The idea of resource redistribution has been around for a long time, but it is not making much headway especially when you are talking about billions of malnourished people.

We could concentrate on preventing crop failures. Crops fail for a myriad of reasons, the most common being, weather, disease, and war. We already do what we can to avoid them, but they are always going to happen.

OK, then, let's eradicate poverty so that people can buy the food surpluses in other parts of the world, converting potood into food. But how do you raise the standard of living for billions of people let alone do it without chewing up whatever is left of our natural resources? There may already be too many of us to do that.

Let's pretend for the sake of argument that there are only a few million people in China, Africa, India, and South America.

Let us also pretend that North America and Europe look the same as they do today. Malnutrition resulting from extreme poverty would be almost nonexistent, help for anyone caught in a famine would be timely, and critical ecosystems would be intact. In short, we would not have a population crisis. We would have a world where the standard of living has been roughly equalized at a very high level—a world with very little poverty. It would also be a world with two-thirds fewer people than today. This is not a proposal to eliminate the people in the third world; it is a thought experiment. I apologize for wasting the time of some readers to clarify this point but there are those who would eagerly misconstrue what I just said.

There is also a growing clean water crisis. Sticking with my moon analogies, Europa, the ice moon of Jupiter may have enough clean drinking water to meet our needs a million times over. The water is not available because it is not economically feasible to distribute it to the population concentrations that need it. I strongly suspect that it would cost billions of dollars to obtain a single cup of water from Europa. It is time for another new word for potential water—*poater*. You can relax; I'm done making up goofy words. Now we have three new terms: poperty, potood, and poater. Desalination plants at the edges of oceans near population concentrations can provide water, meeting the criteria for availability, but the cost would be horrific, making the fresh water unaffordable. It therefore fails the definition for water and remains poater.

With clear evidence that our population is not going to increase forever—a foregone conclusion to anyone with a mind—the debate shifts to the carrying capacity of Earth. To give every person on this planet a standard of living similar to mine or that of the staff of Food First is a physical impossibility. It might be possible to raise every human out of poverty and live in equilibrium with the planet's natural resources with say two billion or so people, but to claim that we can do that with nine billion is more than optimistic, it is naïve, self deceptive, and dangerous.

When people talk about eliminating poverty, some are suggesting the creation of a giant welfare state where the government equitably distributes food and money. That has been tried, it is called communism, it does not work, and everyone knows that by now. To eliminate poverty, you must bring people into the free market. To put it simply, they must have decent paying jobs (or synonymously, business ownership—small farms and so on).

I am now going to give some real life examples of how to reduce poverty and redistribute wealth. I recently decided to buy a tile cutting saw. I went to the local Home Depot and looked at the selection for homeowners. There were two choices in the $100 range. One was in a smallish brightly colored red, white, and blue box. The printing proudly stated that it had been made in America. The other was in a plain brown box. I hefted the two boxes. The plain box weighed at least three times as much. This suggested to me that it might be the best buy. I then opened the boxes to look at the saws. The American saw was made of plastic and had a tiny electric motor. The one in the brown box had no plastic, a huge motor, a stainless steel table, and came with a diamond coated cutting blade. Although the saw I chose to buy had been shipped all the way from China, it was by far the better bargain. How could this be? The answer is simple; the average Chinese wage is only 2.1 percent of the average American wage. China has joined the free market. The Chinese workers are glad to have jobs. They can have more pork with their rice, buy a scooter to replace their bike, and maybe afford a home where they do not have to share a room with four other people. The wages are low now, but they are rising and given enough time, they might match American wages. That is called poverty reduction and it has ramifications. As a consumer, I benefited greatly with this high quality product for such an incredibly low price. However, the American manufacturer will soon be out of business and that company's employees will be at least temporarily unemployed. This is an example of wealth redistribution based on fair competition. My money went to China.

Of course, the increased demand for pork, scooters, cars, and housing, will put even more stress on the planet's resources and ecosystems. The United States consumes about 25 percent of the world's energy and resources. China has a population that is about five times greater than ours. If China's standard of living were to grow to match that in the United States, then the two countries would consume 150 percent of the world's energy and natural resources. Clearly, this is not going to happen. Something will have to change. It would not be very different if the Chinese emulated a Japanese lifestyle instead of an American one.

Let me give you another example of wealth redistribution. People are pouring across America's southern borders by the millions in an attempt to reduce their level of poverty, or inversely, increase their standard of living. This would reduce the population in Mexico, creating more jobs for those who stay behind if Mexico's birth rate were lower. The present Bush administration is pushing legislation to allow all of the illegal immigrants to stay under the auspice that they are just taking the jobs that Americans don't want. Interestingly enough, at the time of this writing, there are 8.2 million unemployed Americans. In reality, this legislation will go a long way to secure the huge Hispanic vote in an election year. It will also fuel the economy in an election year. How? Investors will take advantage of the low labor rates—the result of having millions of poor immigrants—to build things. That is how we built our railroads. It is a way of redistributing the wealth from middle and lower class Americans to even poorer immigrants and makes for greater profits for investors (the wealthy). This makes me sound like another knee jerk power-to-the people activist, which I'm not. Given time, immigrants assimilate and their earning power goes up as an economy expands. But, until that happens, they do make the rich richer, and they do compete for jobs with those who proceeded them. People who make a living in the construction industry are forced to employ these low wage immigrants to stay in business.

It has been the history of our country to have immigrants come in, assimilate, and then move up the economic ladder. Competition with those new immigrants and the descendents of previous ones has always been a source of contention. The swelling population of the U.S. will require more subdivisions, water and sewer services, lumber, and so on which translates into economic growth, poverty reduction, urban sprawl, and ecosystem degradation.

As a person who is obviously convinced that the world would be better off with far fewer people, I am uncharacteristically agnostic, or neutral on the subject of America's out of control immigration policies. I empathize with those trying to improve their lot in life by whatever means available to them. I'd do the same thing if I were in their shoes. Controlled immigration is one of the best ways to redistribute wealth and offset the effects of a population growing top-heavy with the elderly. My concern is primarily the effects of overpopulation on the planet's biodiversity. If the population of Mexico were shrinking because of low birth rates, the picture would look very different. People would still flock to places where they thought they could improve their lot, but at least they would leave behind a chunk of the planet to go fallow and heal. With a growing population, poverty reduction exacts a heavy toll on the planet.

Yet another example of wealth redistribution is for corporations to send jobs overseas—outsourcing. Loyalty to country, state, or employees rarely gets in the way of corporations seeking profit. For example, most of the parts for Boeing's jets are built overseas. In the future, they also intend to offload most of their engineering. As a side note, Airbus, Boeing's only competitor, has recently overtaken them in sales—yet another form of wealth redistribution. Imagine losing out to a consortium of English, French, Spanish and German companies that do not even speak the same languages, or use the same currency, and have their manufacturing facilities spread out all over Europe.

India is also starting to join the world free market. They are not following the same path as China, which is becoming a manufacturing giant. India is becoming a service giant. India is producing large numbers of high quality engineers and computer programmers. American companies are flocking to India to increase their profit margins. The highest paid electrical engineering graduates in India make \$10,000 a year. This is a straightforward example of wealth redistribution. Money that would go into the pockets of American engineering grads is instead going to grads in India. Economists worth their salt will point out that, over time, outsourcing will allow for more profit thus allowing companies to hire more American labor. But, what is rarely mentioned is that before hiring more, the owners and investors will line their own pockets with much of these profits. Humans do not have a built in satiety valve for the seeking of status.

The economic systems available to us fall into a spectrum. At one end, you will find unbridled capitalism and the use of slaves. The Greeks, Romans, Mayans, and most everyone else practiced this as a matter of course throughout human history. As you move toward the middle, you will find *regulated* free markets. This is capitalism with rules in place to limit how badly people with power can abuse those who are making them rich. Anti-trust laws break up companies that have started to swallow all competition. This does not always work. These companies resist and sometimes they succeed, as Microsoft has so far been able to do. There are laws to limit how long you can make your employees work, the conditions they work under, and laws to insure a minimum wage. Workers are called employees at this point, or sometimes wage slaves. Next on the scale comes socialism. Socialism and free market systems begin to blur as taxation and the size of government bureaucracies creep up. Finally, as taxation reaches 100 percent you have a communist system. Experience has shown that a system that sits in the middle somewhere seems to work the best.

I listed these examples of poverty reduction to drive home to those who strive to end poverty without a population reduction the real costs of doing so. Tariffs and legislation to keep jobs at home are also defeating wealth redistribution and poverty reduction. The struggle to eliminate poverty would be less overwhelming if we could just stop our population growth at the lowest possible number and allow it to shrink. Even if we hold at a number like seven billion, we will irreparably damage our ecosystem in very short order.

If, after having said all of this, I have given some the impression that I am against poverty reduction, let me summarize. I am not against poverty reduction. I am against destruction of the planet's remaining biodiversity. More specifically, I am for poverty reduction that does not add to the planet's burden. Population reduction is clearly necessary if we are to successfully reduce poverty *and* save what is left of the planet's biodiversity. I would fully support any idea that could give eight or nine billion people my quality of life and at the same time, preserve all that is left of our natural ecosystems. I just have not seen that idea yet. The only idea that seems to have a reasonable probability for success is to simultaneously reduce poverty and our population while simultaneously protecting what is left of our biodiversity. I cannot envision having success with this many of us striving to stay fed or increase our status every waking hour of the day.

It is unconscionable that we leave our children's grandchildren a world devoid of tigers, river dolphins, and all of the other thousands of species that we share the planet with.

How many of you reading this book have ongoing conflict with one or more of your neighbors, family members, or co-workers? Are you a conservative or a liberal? What are your feelings toward the opposing side? How many of you are Atheist, Moslem, or Christian? Are you pro-life or pro-choice? How warmly do you feel toward those other groups? What is your annual income and what percentage of it do you donate to the less fortunate? How many of you supported the war in Iraq? If you have answered these questions honestly, you should be hard pressed to deny your human nature. Changing it is the only way to change the way we abuse each other and the planet we live on. The problem is that you cannot change human nature because it is part of our genetic heritage. No normal parent will subjugate his or her own child's well being for a stranger's child, and especially not for another species. That instinct is almost as strong as the one that prevents me from putting a gun to my own head as my contribution to population reduction. Human beings cannot help but to argue, compete, make love, and seek status. The planet is dying a death from over six billion cuts.

Our genetic heritage, through the use of chemical rewards, pushes us to seek status, just as it entices us to have sex and compete. What's this? You aren't into status seeking? You have sex because you want to? Competing is below you? Shake off the self-deception. You have little choice in these matters. Does this mean that we are to be exonerated as helpless prisoners of our human nature? The question is nonsensical. Exonerated by what?

There is no simple answer. The TIFIC should be viewed only as a relief valve. It would be only one part of the solution. By reducing population pressures, we stand a better chance of saving our planet's biodiversity and minimizing hunger as we go about the business of being human.

It is time for those who are deeply concerned about population issues to accept that the bulk of humanity is going to remain fed for the indefinite future. It is time to stop emphasizing the potential for future mass famine and start concentrating on the here and now. Let those who *don't* want to believe the world is overpopulated continue to staff and run relief agencies. Let them take responsibility for snuffing famine whenever it raises its ugly head. Those who strive to end social injustice and poverty must continue to do so. More power to them *as long as their plans don't call for further increases in population*. Fortunately for all causes, dropping fertility rates are proof that all we need to do is continue to provide people with education, the choice, and the means.

It is time for us to concentrate on the undeniable fact that we are destroying the biodiversity of the planet and that the root cause of that destruction stems from the wants and needs of billions and billions of people. Gambling that humankind can survive in a world devoid of biodiversity is not a risk we can take. It is essential that we strive to preserve what is left while emphasizing the need to reduce our population. There will be nothing left to protect if we don't. We must not lose sight of the core issue as most environmentalists today have done. We must congratulate ourselves for our success at lowering fertility rates, continue to fight for freedom of choice, and prepare to successfully fight the coming battle to keep patriotic and pronatalist policies from exhorting citizens to have more children instead of sharing wealth through properly regulated immigration.

CHAPTER 5

Lizard Hill

I had just finished a two-year stint as lead design engineer for the shear-tied wing ribs and engine nacelle support structure on Boeing's new 777 airliner. This was the first airliner in history to be engineered entirely on computers and without a physical mock-up to verify that all of the parts would fit. Every part was modeled as a Boolean solid in three dimensions and moved into place with other solid models to insure everything interfaced properly before the engineering was released. The young engineers responsible for pulling this off gave it everything they had. We worked seven days a week, 10 or 12 hours a day, and sometimes all night to make our release schedules. We were hell-bent to prove what we could do. It was an unparalleled success, but some of us paid a price.

A few weeks after our last engineering release, I visited my dentist for the first time in two years. He found an abscess on my gum that had resulted from a botched root canal a few years earlier. I was scheduled for surgery. The next day I saw my dermatologist. He found a large skin cancer on my back. I was scheduled for surgery again. A few days later, I saw an internist for my stomach pains. Apparently, two years of washing aspirin down with coffee had taken a toll on my stomach lining, allowing the H. pylori stomach ulcer bacteria to take hold.

I needed a break. My wife and I started looking for vacation property. Waterfront sounded appealing, so we started there. We also wanted this property to be reasonably close by, so we limited our search to two nearby counties. We had embarked on a steep learning curve. We looked at property on one lake after another. Some of the lakes turned out to be seasonal, drying up in late summer. We learned to ask about that little detail because the real estate agents didn't usually volunteer the information. As I stood at the end of one small lake, listening to the chainsaws and dogs, watching smoke waft up from the burning piles of brush, I realized that this lake, like most of the others, had been ruined by development. It was completely ringed with houses, each with a septic system and well, a lawn, cats, and dogs. It was a small lake—longer than it was wide. Instead of looking across the street at your neighbor, you looked across a narrow body of water. What was the point? We looked at one last lake. As I stood in the wooded lot, thinking that this one might be acceptable, a neighbor's dog sauntered up, sniffed my shoe and proceeded to deposit a steaming pile of shit on the path I had just walked down. No, what I needed was solitude. To hell with waterfront property, I did not want to participate in the rape of another lake anyway.

The criteria had suddenly changed. Now we asked our real estate agent to find a large piece of remote and isolated property within hiking distance to an undeveloped lake.

Our agent called us a few weeks later. She had found fifty acres of undeveloped land that was being sold in five-acre parcels. The property was bordered on one side by state forest and on the other two sides by forest owned by a Boy Scout troop. Access was along a rough logging road, and there were five small, totally undeveloped lakes within hiking distance. It sounded too good to be true. I followed my agent to the property on the following day. There was only one parcel left in the far corner of the fifty acres. It was just what I was looking for. Even if the other lots were eventually developed, we would at least have forest to our back.

I learned that this land had been used as an ad-hoc Christmas tree farm since the 1940's. Most of the big trees had been harvested before putting the property up for sale. This is standard operating procedure—take your lumber profit and turn the clear cut over to developers. Fortunately, the landowner had left enough large trees behind to make it marketable. You can get more money for land if there are some big trees on it. This was also being touted as view property. By clearing most of the big trees, beautiful views of the mountains had temporarily opened.

The acreage had been surveyed, complete with cul-de-sac turnarounds. Everything was poised to convert it into a subdivision.

In Mason County you need three things before you can build on land without a community water or sewer system. First you must pay to have a well dug. This can be an expensive and risky adventure. The well has to be able to produce a minimum amount of clear drinking water per hour. If your property passes that hurdle, you now have to do a perc-test. For this, you must hire someone with a backhoe to dig a ditch to see if the soil will allow sewage water to percolate adequately. The last rule is that your house must be connected to the county electrical grid. You can't build a house or cabin and expect to power it with wind, solar, or propane. This is a good way to slow the proliferation of cabins and shacks in forests. It was interesting to note that the real estate agent made no mention of these minor details. Fortunately, I did not intend to develop this land. In fact, the building code stipulations only reinforced my decision to buy. I knew that the nearest power line was four miles away down a logging road. The cost of bringing it to our land would be prohibitive. None of this property was going to be developed anytime soon.

One day, while standing on a knoll above our property, my young daughter caught a western fence lizard. Later in the day, she saw an alligator lizard. When my wife asked her what we should call our property, she had replied "Lizard Hill."

All was not well in paradise. A guy named George had purchased the first lot. He had placed a hand carved totem pole at the corner of his lot that read, "George's Coner." He had left the letter "r" out of the word "corner" by accident. This was a sign of more to come. The next thing he did was park a trailer on his property, then another, and another. Before long, car hulks and old tires began to pile up. He stole his water from the Boy Scout's water tap and God only knew what he was doing with his sewage. Outhouses are illegal nowadays. Finally, the board of health forced him to put a Porta-Potty on his property. Within a year there were four or five dilapidated trailers and several junk cars along with a continuous entourage of drunken losers hanging around. The cake was iced with a breeding population of pit bulls that threatened and bit passers-by—one of them being my wife. As they would chase my car, I took great pleasure in opening my door and hearing a "thunk" as I nailed the dogs that hadn't learned better. On a positive note, George provided further insurance that no one would invest a lot of money developing the area.

I put a small RV camper on our property. Fully self-contained, these campers come complete with a bathroom, a kitchen, and comfortable beds. We would bring fresh water to add to the storage tank with every trip out. We also hauled our sewage away to the nearest trailer sewer hookup using containers designed just for that purpose. These campers are a big improvement over the old camp stove, outhouse, sleeping bag, and tent arrangement. They're also very eco-friendly when compared to the alternatives I just mentioned. Once the trailer is hauled away, it leaves no footprint, no foundation, and no septic system behind. I hired a guy with a pickup truck to haul it out to the property for me, but he could only take it as far as the logging road. To get the trailer up the hill and deeper into the woods I hired someone with a bulldozer. I was to meet him at the property a week later.

As I approached Lizard Hill to meet the dozer operator, the first hint that something was wrong came to me as the powerful smell of pinesap. The next hint was smoke. Finally, as I rounded the last turn, I found to my horror that the forest south of me had been clear-cut. The death of thousands of fir trees creates a smell that is inappropriately pleasant. The smoke was from the piles of slash being burned. We dragged the trailer to the secluded spot I had chosen which was now completely exposed. I paid the operator and thanked him. He left me standing there gazing over the smoldering landscape. It was later explained to me that the Boy Scouts had needed some cash, and although most of the trees were not big enough for lumber, they were just the right size to be used for paper pulp. I may be writing on paper made from those trees at this moment.

This was the first of many lessons Lizard Hill would teach me. Although my intent was to preserve this piece of property, my urges to improve it were difficult to resist. I brought in a picnic table and benches. I made two platforms out of cedar, one for a tent and the other for a portable hot tub. The lumber, of course, had all come from clear cuts somewhere.

I knew from the start that anything not nailed down would be stolen. I had a security system in place within days of parking the trailer. I started with a simple alarm hooked to the door. Someone dismantled that one fine day and cleaned out all of our canned goods.

Next, using a car cell phone, an intermittent windshield wiper switch, and a small electric motor, I built a system that would call me at home whenever the trailer door was opened. It had a speakerphone so I could hear and talk with anyone in the trailer. This was a novel idea, but the cell signal was unreliable way out in the boonies.

My third system proved to be the best. I decided that it would be smarter to keep trespassers from entering the trailer in the first place. I bought a 12-volt motion sensor rated for outdoor use and hid it about a hundred yards down the logging road. I attached it to a car tape deck hidden in the trailer. I made a recording of televangelists spouting fire and brimstone and put that into the tape deck. As you approached the trailer you could just make out the sound of angry voices. As you got closer, you would hear a deranged man shouting about God and hell. It worked flawlessly for many years. I added to this defense "no trespassing" signs, which claimed that digital motion sensors and surveillance cameras connected to cellular transmitters monitored the property. Although it wasn't true, the signs made potential trespassers nervous.

Over time we learned what kind of animal life we could commonly expect to see. The lakes are full of newts. These newts are aquatic except for a few months in the winter when many of them take on a terrestrial form and go marching off to find new ponds. They can sometimes be seen walking across snow. This is the safest time for a migration because few predators are active in the cold wet winter. My daughters fear that one-day they will come back to these lakes and find no newts. It would break their hearts if I told them that it is almost inevitable, so I don't.

Red-legged frogs can be found on rare occasions. These large frogs are not usually found near water except during mating season. They used to be so common that logging camps would sometimes supplement their meals with frog legs. No one knows why their numbers have declined so drastically. Puget Sound garter snakes are quite common. We once found a skeleton of a garter snake with the skeleton of a newt in its mouth. The newts secrete a poison from their skin. I don't know if the snake was poisoned by the newt or run over by my car while in the act of eating it.

In late winter the ponds fill with the tadpoles of the emerald tree frog. These ponds dry up in the summer, but not until the tadpoles have turned into frogs. Wetlands that dry up are important for many species of frog. Large predators like fish are absent which dramatically decreases predation. Missing from this property was a pond large enough to avoid drying up every summer. Some species of frogs and salamanders require more than one summer to mature and need ponds that can last through the dry season most of the time. I decided to make one. I picked a ravine that could be dammed at one end, and hired an excavator. I then laid a rubber pond liner in the bottom and hired the bulldozer guy to bury the liner. The finished dimensions were ninety feet by forty feet and about ten feet deep at one end. As I was paying the bulldozer operator, I noticed a fatally injured rubber boa lying at my feet. The weight of the dozer had crushed it as it lay in its shallow burrow.

Coyote and deer were once quite common. One night I left the alarm system activated and was awakened on two occasions—once by a deer and once by a coyote. Their numbers have dwindled in proportion to the army of brush pickers that now comb the area on a daily basis.

Western fence lizards can be found as well as western alligator lizards. The former lays eggs and the latter bears live young. We found this out while keeping some as temporary pets.

The large pond nearby always has beavers and occasionally otters. As the years went by, the lots adjacent to me were put back on the market. I purchased five more acres.

One weekend while camping with my family and enjoying the solitude, a group of Spanish-speaking brush pickers swept through the adjacent forest talking and whistling as they went. The next day a group of Vietnamese-speaking pickers did the same.

Brush picking is a relatively new occupation. People who pick brush for a living are looking for parts of wild plants that are used by floral companies to accent their flower arrangements. Using wild greens is the latest fad. Whenever you buy a bouquet keep in mind that those greens were all picked by hand in north-west forests. The pickers bring their bags full of greens to large warehouses where they are sorted and shipped out to the florist shops. It is a growing industry. I have watched the construction of a dozen such warehouses in the past ten years. Much of the greenery is shipped to other parts of the world. Brush pickers stomp back and forth, bushwhacking through forestland over and over again, gleaning specific greens from it—salal, moss, bear-grass, evergreen boughs, ferns, wildflowers, herbs, huckleber-ries, nuts, and mushrooms. The harvest in Washington State alone will exceed $236 million in 2003.[140]

I constantly find evidence on our land that brush pickers have trimmed the lower branches off the young western pine trees. Each year they trim a little higher on the tree until there are almost no branches left. The boughs are made into Christmas wreaths. I once caught a picker pulling the bark off a young cedar tree in long strips to be used for weaving. Authorities recently found over 600 bags of illegally picked moss in Kitsap County in Washington. About three-quarters of the picking is done illegally and paid for with cash. Few taxes are paid.

One of the worst things about brush pickers is that they don't follow trails. In reality, they are on the lookout for anything of value, not just the greens used by florists. Forest Service en-forcement officers spend about half of their time on thefts of non-timber products. As they bushwhack, brush pickers litter, urinate, defecate, and make lots of noise. An enhancement to the ecosystem, they are not.

A couple of professors have been studying the environ-mental impact of brush picking. They haven't come up with anything conclusive yet, but it doesn't take a rocket scientist to know that brush picking isn't good for the environment. As the numbers of brush pickers continually grow and as time marches on, our forestlands are beginning to look like trodden city parks. The privacy needed by wildlife to give birth and raise their

[140] http://seattlepi.nwsource.com/local/58249_greens14.shtml

offspring is being violated 365 days a year. The cedar tree with the stripped bark is dead, my pine trees look sickly, and my picnic table and benches along with anything else that wasn't bolted to the Earth have been stolen. My property is never without the litter and toilet paper left behind by roving bands of brush pickers.

Everybody is making money—the landowners that lease their land for brush picking, the brush warehouse owners, the brush pickers, and the florists. Once a new industry that relies upon natural resources gets entrenched, it becomes harder to remove it. Many people now depend on the industry for their livelihood and florists are accustomed to working the greens into their arrangements. What we need is a federal law making it illegal to use wild picked greens in floral arrangements. It would do no good for a single state to make it illegal, because the demand would still exist and the supply would still meet demand, one way or another. Florists would lose no profit under a federal law because it would be a level playing field. Since no florist would be using wild greens, one could not gain an advantage over the other by having greens to enhance their bouquets. If prices for floral arrangements remained unchanged, the florists' profit margins would actually increase because they would not be paying for the extra greens.

There's the problem of all the people who make a living from the industry. If you simply pull the plug on their livelihood, it will create hardship and resentment. We should all pay with our taxes to reimburse those who were legally making a living in this industry for a transitional period of time giving them a chance to find employment through other means. Buy out the brush shed owners, give the workers a significant severance pay, and lease the forests for a decade. Everybody would be happy and the industry would be gone. It costs to protect the environment. The environment continues to be chewed up by people looking to improve their lives. It is only natural.

Deer mice will completely overrun any cabin, trailer, or house that sits in a rich ecosystem. Deer mice can carry the dangerous Hanta virus, and this was a concern to me. There is no way to keep them out. This explains why people have always kept cats around. I would set five traps every week, and still I couldn't keep up with them. One day I entered the trailer and found that my burgeoning mouse population had attracted two weasels. I never could find out how all these animals were gaining access to our trailer. A mouse can fit through a hole no larger than a dime. All it takes is for one mouse to find such a hole. It will then remember that location. Its mate and any off-spring will also learn of the hole. A heavily used scent trail will then attract other mice to the same hole. That is when the flood-gates open.

Next came the insects. The carpenter ants and termites waged open warfare for the last remnants of the wooden structure. The yellow jackets and hornets also vied for real estate. Out of desperation, I finally resorted to using poison to control these pests.

Finally, there were the human pests. Nothing of value could ever be left behind, or it was stolen. Who were these people and where did they come from? Drifters, hunters, brush pickers, who knows?

This is why I choose to live in a city. I love nature and I love to visit nature, but I do not want to live as one with nature. What is the point of living in the woods? You need to maintain roads to get your car there, keep guard dogs to protect you from human predators, and cats to protect you from rodents. In addition, you must spray insect poison everywhere to protect your property from wood ants, termites and hornets. Your property quickly becomes a blight on the skin of the Earth. It is wiser to live in town and visit the woods.

I have watched the construction of two homes within a ten-mile radius of this property in the few years that I have owned it. The trees are cut down, the bulldozers come in, and a septic system and well are installed. After the house is built, the cats and dogs arrive. These homes, almost without exception, are on the market a few years later to begin an endless procession of turnovers. The dream of living in the woods usually fails to meet expectations. It just isn't worth the hassle. It's kind of like owning a boat. We have all heard the cliché "The two happiest days of boat ownership are the day you buy the boat and the day you sell it." This is also often true of vacation property and homes built in the middle of nowhere. Even George of "George's

Coner" finally disappeared ten years after his arrival—leaving behind one hell of a mess. If our population were not constantly expanding, these homes would never have been built.

One of the lakes in the area was purchased by a guy named Larry for about half a million dollars. He immediately built a large brick bed-and-breakfast mansion on the edge of it.[141] This was a serious case of urban sprawl. He is, however, a good steward of his land and continues to allow public access to one part of the lake—catch and release fly-fishing only. He keeps this lake well stocked and otters sometimes take up residence there. They cost him money because they eat the stocked trout. This is an example of the price of protecting the environment. The government will issue him a permit to shoot the otters because they are damaging him financially. I would rather see the government reimburse him for his losses with the tax dollars we all pay rather than issue a permit to kill the otters.

Lizard Hill taught me a lot about conservation and most importantly, a lot about human nature. If a person like me, who loves nature as much as I do, has a hard time controlling his urge to improve his property (wreck it), how can we expect the rest of humanity to resist the urge to control their own environment?

It also taught me that I am mortal. There is a limit on how much time and energy an individual can invest in vacation property. A trip to Lizard Hill required that I bring out the tub heater, two freshly charged car batteries, and twenty gallons of fresh water. That was on top of everything else such as food and tools. Occasionally the septic holding tank needed to be emptied into a portable tank so I could transport it to a trailer dump station. The propane tanks often had to be taken off and refilled. One day I went to Lizard Hill to prepare it for a weekend of camping with my family. I found that the hot tub had a small leak. The tub heater was also malfunctioning. The alarm was on the blink, and consequently someone had stolen one of my batteries. The trailer was infested with mice, termites, yellow jackets, and wood ants. When I tried to start the furnace a mouse nest caught fire. As the trailer filled with smoke, I feared starting a forest fire because the summer had been very hot and dry. I managed to put the fire out and stood there in silence. I heard a

---

[141]http://www.cadylake.com/

strange clicking sound coming from the trailer walls—the sound made by warrior termites when their nest is disturbed. They gnash their mandibles together as a warning to intruders. The smoke had upset them. It had been a good ride. I gave the trailer away a few weeks later to someone who planned to live in it until he finished building his home. Wisely, he was going to fumigate it first.

I have to put in a plug here for my 1989 Jeep Cherokee. Feel free to skip this part if you aren't a guy. The recipient of my moth eaten trailer had come to retrieve it with a full-size pickup truck—the kind with four wheels in the back. He couldn't get the trailer up the steep logging roads. As you may recall, I had hired a bulldozer to get it where it was. He finally suggested that we hook my Cherokee to it. I was skeptical. He told me that he had seen Cherokees move mountains. Unfortunately, my Jeep was trapped behind the trailer on the only logging road out. About this time, it started to pour. I knew of another trail, but it crossed very close to my pond on a steep incline. I opened my windows in case the Jeep rolled into the pond and then gunned it through the narrow passage, two wheels in the water—no problem. I put the Jeep in four-wheel granny gear and backed it to the trailer. We hitched it up and I tentatively put the Jeep in forward. The Jeep didn't seem to notice that a 30-foot long trailer was behind it. I pulled it for three miles to the paved road. I guess I hadn't needed the bulldozer to move the trailer when I first brought it to Lizard Hill.

I might as well take this opportunity to tell my other Jeep story, again, feel free to skip it also. It was January, and I had just taken the ferry from Seattle to the Kitsap peninsula to start my 45-minute drive to Lizard Hill. As I rolled off the ferry, two feet of freshly fallen snow greeted me. I decided to see how close I could get to Lizard Hill. I stopped at an auto parts store for some tire chains. The man at the counter told me that a Jeep with chains is unstoppable. I thought that unlikely. Some of the main roads had been plowed. I reached the turnoff for the last five miles of pavement before the logging road that led to Lizard Hill. I stopped and looked at the snowfield where the road should have been, bordered on both sides by forest. I slowly drove the Jeep onto the snow ready to back out at the first sign of trouble.

Snow piled up on the front bumper and onto the hood. I kept an eye on the temperature gauge because the radiator was also covered with snow. As the engine temperature climbed, I would stop to clear the snow away. I had to drive around a few abandoned four-wheel-drive vehicles that had tried to enter this road from side-roads. So far I was the only vehicle to successfully navigate it. Finally, I came to my turnoff. I stopped and looked back at my lone tracks winding off into the distance. I turned up the logging road and immediately got stuck. Cherokees do have their limits. I put some branches under my tires and managed to turn the Jeep around, pointing it back the way I had just come, just off the main road. I hadn't gotten into trouble yet so I decided to hike in to the trailer, which was three miles down the logging road. About three-quarters of the way there I realized I was finally in trouble. Hiking in thigh-deep snow was proving to be more difficult than I realized. I decided that the safest thing to do was continue on to the trailer were I could rest and get warm. Once I reached it, I wasn't sure I had the strength to walk back out. I didn't like the idea of spending the night or of leaving my Jeep in the middle of nowhere. I felt better after an hour of rest. I assumed it would be easier on the way out because I could walk in the tracks I made coming in. I was approaching the limit of my endurance by the time I made it back. The sun was just setting. I looked up the paved road and saw a snow plow coming.

When it reached me the driver stopped and rolled his window down, "I've been following your tracks for five miles," he yelled. "My next car is gonna be a Cherokee."

Too exhausted to do anything else, I gave him a thumbs-up. There is an old saying, "Four-wheel drive should only be used to stay out of trouble; it should never be used to go looking for it."

It wouldn't surprise me if some of you are appalled that I drive an SUV. I bought it many years ago to get to Lizard hill via rough logging roads. I also use it extensively for my business, which involves hauling heavy tools and pulling a trailer. It may be the only Jeep in Seattle actually used for what it was designed to do. I have recently looked into replacing it and have concluded that it is by far the most efficient vehicle available for the task. There is no question that it also brought me a measure of status when it was new and shiny. Give me a break, I owned a series of four Ford Pintos prior to buying this used Jeep. When I was eighteen I did a painting titled "My rotary engined, hydrogen burning, Pinto car" that won a gold medal in a citywide art contest. The painting was done at the height of the gasoline shortage caused by the OPEC embargo. The idea of using hydrogen to fuel our cars has been around for a long time.

It has been a full decade since we bought that property. The trees in the clear cut are ten feet tall now. We had many wonderful times there. Some day, somebody will bring a power line in and that will be the end of Lizard Hill.

Before I finished this essay someone had replaced George and his dogs, and I found yet another abandoned pick-up truck on my property just today. As I rode back on the ferry, the headlines in the local paper read "New zoning laws allow more homes to be built in rural forests." A battle is looming to prevent a local lumber family from building a 700 home subdivision in the middle of their forest holdings, just a few miles from Lizard hill. It will never end until our population stops growing and even then it will not end if our personal wealth continues to grow such that we will desire vacation property. The forests need to be roped off by using free market methods similar to those discussed in other parts of the book and protected from development and fragmentation. Zoning laws and government ownership do not present a long-term solution.

CHAPTER 6

Environmentalism American style

This chapter is not meant to advocate lock-step support for the latest environmental party line. In fact, it takes a critical look at some of our activities. It cannot be possible that every popular environmental mantra is the optimal one. Conversely, it is not possible that everything anti-environmentalists say can be wrong. They get it right, usually for the wrong reasons, once in a while.

Global warming

I have learned from experience that to suggest that we may not be successful at containing global warming through world-wide emissions legislation is to invoke a knee-jerk reaction in many environmentalists. That suggestion got me labeled as a monomaniac in one instance and an anti-environmentalist in sheep's clothing in another. Should we pour billions of dollars into roping off preserves and reducing our numbers, or should we spend billions trying to reduce CO_2 emissions with air pollution regulations? A basic tenet used by physicians is to treat the cause of an illness, not its symptoms. Attempts to reduce CO_2 levels are treating the symptoms of what ails our planet, not the cause. The cause is overpopulation—the needs, and desires of billions of people. The number of cars being built goes up every year. You get nowhere by doubling gas mileage if you have twice as many cars. Giving aspirin to reduce the discomfort of a fever—global warming—is not as effective as prescribing an antibiotic—the TIFIC— that would reduce the number of agents that caused the fever—us.

First off, forests eat CO_2. Increasing forest cover would decrease CO_2 in the atmosphere. This idea has been proposed of course, but research has suggested that it would not be as cost effective as reducing worldwide emissions—assuming one can reduce worldwide emissions. Some who favor emissions legislation have actually proposed that trees increase global warming. Apparently, they act as solar collectors by darkening the ground. Why is it that we human beings are incapable of reaching consensus on anything? This is ridiculous.

Secondly, if there were two billion of us instead of 6.35 billion, our air pollution concerns would disappear on their own. Why not view population reduction as the most efficient way to reduce emissions, just as administering an antibiotic would be the best way to treat a bacterial infection that is causing a fever? In other words, I am not against limiting CO_2 emissions. I am saying that we are going about it in the wrong way. Anticipating a counter-argument, one might say that accomplishing those two goals of reducing our population and increasing our forest cover is a dream. I would counter by saying, read this book, it is one long argument to the contrary. Attempting to control the climatic swings of an entire planet by imposing worldwide emissions standards on nine billion human beings is what strikes me as being unrealistic. That idea takes at least as much hubris as the idea that we can reduce our population.

I strongly suspect we are politically incapable of reducing emissions enough to make a difference. China is becoming a major manufacturer. What are the odds that we can reduce world emissions 50 percent when our population is slated to increase by 50 percent? How can we ask billions of people striving to pull themselves out of poverty to not drive cars or scooters once they can afford them?

Let's say, for the sake of argument, that we successfully rally the countries of the world around our belief that we must reduce emissions to stop global warming. Then imagine that we succeed in reducing emissions, and global warming consequently stops. Here is the problem: warming and cooling trends are a fact of life on this planet.[142] There is no such thing as climatic equilibrium. Let us assume that the Earth then starts to move toward a

---

[142] http://www.csmonitor.com/durable/2000/08/24/fp16s2-csm.shtml

cooling phase on its way to another ice age as it has repeatedly done in the past. Will we then try to rally the nations of the Earth to increase CO_2 emissions in an attempt to stave off a cooling period? Will we lobby to cut down the Amazon to prevent it from absorbing too much CO_2? If I am wrong, and we are able to reduce CO_2 levels enough to reverse global warming, how long will it take? With our population heading for nine billion, time is of the essence. If it takes a hundred years to stop and reverse it, will it be too little too late? What will remain of our biodiversity?

We cannot predict the weather, let alone control it. Back in the fifties, there was high hope that we would someday be able to forecast the weather with ever-increasing accuracy. We have since learned that the Earth's weather resembles a chaotic system. It is not and never will be very predictable. There have been no major advances in predicting the weather for the last few decades. If predicting the weather for five days is nearly impossible, predicting global warming or cooling over the next couple of centuries is especially suspect. I think humanity will adapt to a warmer climate. The problem is that we will be leaving a world stripped of its biodiversity to our children because ultimately, that is what global warming is going to do.

In the short run, we can't predict that global warming will be a net improvement or a net degradation for human cultures. One thing all researchers seem to agree on is that we have staved off the next cooling cycle. We are avoiding the next ice age. With further warming, some parts of the world will get wetter, others dryer. It could all happen so slowly that there might be marginal *net* negative impact to human economies. Admittedly, if we can stop global warming, we should, because it most certainly holds the potential to cause great harm to many people by causing drought in some parts of the world and displacement due to coastal flooding in addition to wiping out what is left of our biodiversity.

Global warming may turn out to be that major planetary event that environmentalists and population activists have been so concerned about, that link in the chain of life that finally breaks. If global warming will mean the end of much of our remaining biodiversity, then the unraveling of our ecosystems will accelerate. Something is making the frogs of the world disappear already.[143]

One thing is certain, any change in the global weather system, whether it is a cooling or warming trend, and whether it is caused by us or not, will have a devastating effect on what is left of our natural ecosystems, as they are so severely stressed by humanity already. For example, if the forest habitat of the mountain gorilla shrinks because of a lack of rain, that will be the end of mountain gorillas. In the past, pockets of habitat would have survived in other parts of Africa where the rain patterns were still adequate. A more striking example is that the retreating ice sheets in the arctic are having a serious effect on polar bear populations. Without the pressure we put on them and their habitats, they would have a better chance to survive a warming cycle. They have weathered such cycles before when caused by giant volcanoes or continent sized forest fires, but that may not be true today. Before humanity overran the planet, a global warming or cooling trend would have caused the ranges of animal populations to be redistributed. Some ranges would get bigger and some would get smaller. Today, nature is trapped in tiny, ever-shrinking pockets. There is no place to run, no place to hide. Global warming will be the straw that breaks the camel's back as far as the last remaining pockets of natural ecosystems and wildlife are concerned.

Most of the coming extinctions will follow the pattern of the heath hen. Once common in the eastern United States, by 1908 its population was reduced to fifty birds by the combination of hunting and habitat loss from human population growth. A refuge was established for them on the island of Martha's Vineyard. This protected group of birds grew in number to a couple of thousand by 1915. A forest fire, a hard winter, and poultry disease, reduced their numbers all the way down to thirteen birds. Inbreeding which results in reduced genetic variation made them especially susceptible to disease. A few years later they were extinct.[144]

---

[143] http://en.wikipedia.org/wiki/Decline_in_frog_populations

[144] The Sixth Extinction by Richard Leakey and Roger Lewin, page 61.

Activism

The environmental movement in the U.S. is fraught with conflict. The radical environmental group calling itself the Earth Liberation Front—ELF—was recently labeled the number one terrorist organization in America by the current administration. Although their actions have caused a great deal of property damage, I cannot believe anybody has ever been terrorized by their behavior. One has to wonder how a group that has deliberately not caused bodily harm to anyone deserves a label like that, but politics are politics. Calling them the number one vandals or arsons in America might be more appropriate.

In reality, there is no organization called the ELF. There is only a website and people who vandalize things in the name of the ELF. I chuckle every time I think about the number one terrorist organization in America having its own website.[145] That FBI sure is crafty. I suppose they expect to nab some members who are foolish enough to visit it—and come to think or it, they probably will. Anyone who wants to vandalize or arson something under the auspice of protecting the environment simply has to leave a note behind claiming that it was done by the ELF. If it gets reported in the news, someone adds that particular act of vandalism to the list on the ELF website. One problem with an organization that has no organization is that there is absolutely no control over what some idiot might do in your name. So far, nobody has been hurt, but an accident will happen someday.

Be forewarned that the FBI is undoubtedly monitoring visitors to this site and adding them to a database. If an SUV gets spray painted in your neighborhood, you may find that you have inadvertently added yourself to the list of suspects. American citizens do not have the same protections from their government that they had prior to 9/11. You might want to think about it before visiting that site. As a matter of fact, right here in Washington State, $50,000 has been added to the Senate Republicans' 2004 budget to make a database of people and organizations thought to be involved in eco-terrorism. Having visited the ELF sight on numerous occasions to write this chapter, I have undoubtedly added my name to that database.

---

[145] http://earthliberationfront.com/doa/2003.shtml

Admittedly, their poorly thought-out and destructive actions have never failed to be counterproductive. They have single-handedly done more damage to the image of environmentalism than any other group I can think of. Although I can emphasize with their frustration, I do not buy their methodologies. When you think about it, this is analogous to the Indian wars at the turn of the last century were the ultimate and final struggle against urban sprawl played itself out resulting in the destruction of all Native American cultures at the hands of others with more money and power.

Typically, the ELF set fire to new construction that is causing urban sprawl. Lately, vandalizing SUVs has become very popular. Their reasoning can be hard to follow. For example, here in Seattle they chose to arson the Center for Urban Horticulture at the University of Washington, a lab that was conducting genetic research on trees, leaving the researchers—most of them staunch environmentalists—stunned and confused. The ELF is closely allied with ALF, the Animal Liberation Front, who just this summer vandalized a mink farm in Washington State by releasing 10,000 of the little devils into the surrounding countryside.[146]

Circumstances were reversed recently when a California court awarded a multimillion-dollar settlement to Darryl Cherney and the late Judi Bari—members of another environmental group called Earth First! This organization splintered off of the ELF and uses less radical means such as tree sitting and protests.

They were driving to a protest site when a bomb went off in their car. The FBI arrested them in their hospital rooms claiming they were victims of their own bomb. Judi, a mother of three, died of breast cancer before the appeals ended.[147]

All movements have similar problems. The anti-abortion forces suffered a terrific blow to their public image when certain individuals decided to bomb health clinics and even assassinate physicians. The problem was exacerbated when some of their leadership failed to denounce these senseless and violent actions.

---

[146]http://lists.envirolink.org/pipermail/ar-news/Week-of-Mon-20030825/005415.html

[147] http://www.judibari.org/

What do you get when you mix hot blooded, adventure seeking young men, semi-automatic weapons, and a self-righteous cause? You get Beirut, Mogadishu, Northern Ireland, and Palestine. The list could go on forever. Violence is a tool of last resort. It was necessary against the Axis powers in world war II, but it has no place within the environmental movement of the United States and Europe.

The problem lies principally in the opposing parties' inability or unwillingness to empathize with each other. Self-righteousness is a dangerous thing. If the environment and all species of wildlife are to be protected for future generations, there is a price tag for doing so. That is easy to say, but much harder to do. It is easy for an environmentalist like me sitting in front of a computer filled with toxic heavy metals to say; it is hard for a lumberjack supporting a family in rural America to do.

Tree hugging

Did you know that there are more trees in America now than there were when the pilgrims landed? It's true, sort of. The curator at a logging industry museum told me this. These extra trees, however, are an assortment of seedlings, saplings, and other small immature trees. The statement is deliberately misleading. When one considers that only 5 percent of our original forest cover remains, anyone with a brain must suspect that something is amiss. The original old growth forests were used for a combination of wood construction and fuel. When wood became scarce enough, coal replaced it as a fuel.

There may have been fewer trees, but there was a lot more wood in this country when the Pilgrims landed than there is today. Trees are just one piece of a huge puzzle called an ecosystem. Logging is just dandy when properly executed. The logging industry does not need propaganda. Those trees were cut down because people wanted the products made from them. The logging industry is just the messenger.

I have done many hikes in the Olympic National Forest of Washington State. The trees are big and the scenery is beautiful, but wildlife sightings are rare. A park ranger admitted to me one day that there just isn't much wildlife there. The large trees in an old growth forest suck up all the sunlight allowing little energy to get down to ground level. Fresh timber cuts, on the other hand, generate a huge diversity of plant and animal life. Just the same, old growth forests are an important part of an ecosystem, and the few that remain should be preserved.

When I was a teenager in Indiana, my mother built a deck using clear, vertical grain, 2 x 12 redwood boards. It was no big deal. I went to a lumberyard just a few months ago and asked to buy a 2 x 12 redwood board without any knots. I was politely informed that lumber fitting that description doesn't exist anymore.

The quality of lumber one finds today is far inferior to that of just 30 years ago. Lumber is coming from small immature trees and this means that it has a lot of adverse grain flow and knots. Lumber yards now stock a huge variety of what is called alternative forest products. Some boards are made of recycled plastics mixed with sawdust while other products are a mixture of sawdust and special resins. Treated lumber—lumber saturated with poisons—is a cost-effective alternative to the very expensive rot-resistant cedars and redwoods. Our highly innovative free market system continues to find ways to stretch our wood resources. It is obvious by the large variety of types and ever growing quantities of alternative wood products available in any lumberyard that we have locally outstripped tree production. There is no other explanation for the appearance of such products. Free market forces don’t create new products for the fun of it. There is clearly a supply and demand issue driving these innovations. This is an excellent example supporting Simon's philosophy that humanity will find ways to stretch dwindling resources, although there is a limit to such a philosophy.

Thousands of logging communities across the United States have suffered the trauma of having no more trees to cut down. Even Seattle was once a booming logging town. These communities sometimes disappear from the map while others adjust. It is usually a long and drawn-out process. The old growth trees go first followed by second and third growth trees. The distances traveled to harvest and bring the lumber to the local sawmills get longer and longer. Eventually the lumber being cut is of poor quality and sawmills begin to close one by one. Finally the trees being cut are not suitable for lumber and are used instead for paper pulp. The last sawmill closes without fanfare or protest because there is no one to blame.

On the other hand, when a logging community suddenly finds itself cut off from its livelihood by a piece of federal legislation meant to protect some tract of forest, there is someone to blame. The affected people band together, write letters to their congressmen, build websites, and organize tractor convoy protests to the Capitol. In these cases, the individuals in that community should be compensated. They should not be asked to bear the brunt of protecting the environment alone. The economic penalty must be spread out and paid for by tax dollars.

For example, maybe people adversely affected by such legislation should receive significant assistance for some extended length of time. In fact, the compensation should be generous to the point that logging communities would solicit the government to protect their forests so that they can cash in on the compensation. Rather like bidding to host the Olympic games. I am not calling for increased taxation. I am only suggesting tax redistribution to protect people when their livelihood is disrupted by environmental protection activities. The details of such redistribution would be for the pork barrel politicians to fight out.

Protecting forests causes lumber shortages. Lumber shortages generate high prices. High prices reduce the number of affordable homes built. There *is* a price to pay to protect the environment.

Having said all of this, the question remains—are we running out of lumber? Not yet, an acre of old growth is cut in Canada about every minute. There's lots of hardwood in the Congo, and lets not forget about all of the softwood in Siberia—where there are only a few hundred Siberian tigers left. With a growing population, the old growth stands in the U.S. that have been protected to date will eventually fall to economic pressures as well. The high price and low quality of lumber today strongly suggests that, at least here in the U.S., we are presently cutting wood faster than we are growing it.

While visiting my local lumberyard during the summer of 2003, I saw a brochure sitting on the counter. It was entitled, "Lumber Suppliers Speak Out on Rising Prices." Customers had been complaining about the sudden and dramatic rise in lumber prices. Four major suppliers had submitted short press releases to explain the problem. The reasons included: the biggest government purchase in ten years (don't ask me what that means), limited logging due to extreme fire danger, low interest rates that have caused an increase in new housing, and tropical storms on the East coast. Oddly enough, there was no mention of the real cause of high prices—the shortage caused by the Bush administration's tariff on Canadian lumber. The high price consumers are paying for lumber is the direct result of the Bush administration protecting forest industry special interests by keeping less expensive, better quality Canadian lumber out of our market with a tariff. The American timber industry is lining its pockets rather than let that money go to Canadian companies and give American consumers the savings that would result from lower lumber prices.

Modern homes are big and consume a prodigious amount of lumber. There are going to be many trees cut in the next few years. The lumber suppliers should see a big profit spike.

This touches on the subject of forest fires, fire prevention, and controlled burning. By controlling forest fires we have created forests filled with upwards of a thousand trees per acre where normally there would only be a few hundred large trees because naturally occurring fires traditionally thinned the small ones out. When these forests catch fire they burn fiercely and destroy even the big trees that normally would survive. They argue that it is not economically feasible to go into a forest and thin the small trees—and in that, they are right. That is why the Forest service is going to allow thinning of trees that are just shy of three feet in diameter to pay for the thinning.[148] The problem is that most of our original forests are gone. Allowing what is left to burn would have unanticipated ramifications—another example of how we have gotten ourselves into a corner. The compromise is to allow lumber companies to go in to cut profitable trees and while they are at it, thin the small ones out. This is much more expensive than just clear-cutting a chunk of forest. The result is higher lumber prices.

---

[148] http://www.forestsfuture.fs.fed.us/pdfs/trees.pdf

As I sit here writing, it is April 22, 2004, Earth Day. I just received e-mail from the Northwest Old Growth Campaign. I thought I'd pass it along:

> More than 150 ancient forest timber sales could now move ahead because of new Bush policies. Several of these sales are currently being logged or are under imminent danger of being logged. Here's a sample:
>
> 1) Cowcatcher Timber Sale, Roseburg Bureau of Land Management (BLM) - currently being logged
> 2) Mr. Wilson Timber Sale, Medford BLM- currently being logged
> 3) Bear Penn, Medford BLM- currently being logged
> 4) Soukow Timber Sale, Medford BLM - currently being logged
> 5) East Five Mile Timber Sale, Mt. Hood National Forest - currently being logged
> 6) Papa Cow, Medford BLM - currently being logged
> 7) Hilynx Timber Sale, Mt. Hood National Forest
> 8) Solo Timber Sale, Mt. Hood National Forest
> 9) Borg Timber Sale, Mt. Hood National Forest
> 10) Pryor Timber Sale, Willamette National Forest
> 11) Straw Devil Timber Sale, Willamette National Forest
> 12) East Devil Timber Sale, Willamette National Forest
> 13) Peanuts Timber Sale, Umpqua National Forest
> 14) Pigout Timber Sale, Umpqua National Forest
> 15) Biscuit Salvage, Siskiyou National Forest
> 16) Knob Timber Sale, Klamath National Forest
>
> The cathedral forests of the Pacific Northwest are a national treasure for all Americans. It's time for the Pacific Northwest Congressional delegation, representing the people of the Northwest and its national treasures, to step up their leadership and challenge the Bush Administration's backwards policies. The people of this nation, expert scientists and natural resource economists are all calling for an end to ancient forest logging.

When you step back and look at all of these attempts to manipulate nature, it is obvious that we are consuming our planet.

In the long run, the only way out is to use our brains to create technology that will allow us to dodge unintended pregnancies, allowing us to reduce our numbers and thus the demand for natural resources. Humanity can be likened to a forest fire, consuming its source of fuel. Today, we are witnessing the crowning of that fire.

Strange bedfellows

The environmental movement in the United States has a propensity to fragment. There are those who are mainly interested in the quality of our air, water, and food. They oppose genetically modified foods, push for organically grown crops, and support efforts to improve air and water quality. They are primarily interested in the human side of the equation, themselves in particular. There are those who are mostly interested in protecting wildlife. Usually called conservationists, this group includes, among others, hunters, trappers, and sport fishermen. Again, many of them have an immediate vested interest in preserving the environment for their own recreational uses. People interested in preserving the planet's biodiversity for its own sake need a unique name to differentiate themselves from the generic environmentalists and the conservationists. I'd like to propose one and call this subgroup, biodiversity preservationists, or even better, biodiversivists.[149] That's a mouthful, but it has a nice ring to it.

The American environmental movement can also form some interesting alliances. For example, the animal rights activists have intertwined themselves with the environmental movement. They sometimes share the same Web pages although the two movements have very little in common. I have seen a self-righteous bumper sticker that says, "If you eat meat, you can't call yourself an environmentalist." Hard core animal rights activists usually have a vegan or vegetarian diet, so, along with worthwhile efforts for more humane treatment of lab and domesticated animals—cows, pigs, chickens, cats and dogs—some would also like to make it illegal to consume animal protein. So now you have animal rights and veganism intertwined with

---

[149] Thanks to Jeremy M. Harris for that one.

environmentalism. I gave up vegetarianism when I realized that converting the human diet to one of grain and beans would make no measurable impact on world hunger or the preservation of biodiversity. India is the example supporting my opinion. The people of India subsist primarily on grain and vegetables, yet they have decimated their natural environment—more on that later.

An estimated 60 million bison once roamed the North American continent. Add to that the number of deer, antelope, moose, and elk, that were also roaming the forests and plains, and you would find that the herbivore and deer population in America was not far off from our current domestic cattle population of about 96 million. We almost drove the bison to extinction and then converted their prairies into farmland. In other words, the resources being consumed by cattle today are roughly the same as that which was consumed by the combined numbers of the American bison and other wild herbivores when they were in their natural state. Much of our food is grown on the prairies that once supported those herds. The planet has long supported vast herds of herbivores, much as the African savanna does today. In the U.S. we have replaced the wild ones with domesticated versions. The problem is that the planet has never supported vast herds of herbivores *and* primates.

A vegan is a vegetarian whose diet consists of plant products only, no eggs, no fish, and no dairy. There is more to it than just the food they eat. It has a philosophy of respecting animal life—but out of necessity, not plant life—similar to some religions in India. Veganism looks very much like a religion except it does not involve deity worship. Veganism is not an attempt to return to nature. There are no vegan hunter-gatherer societies or archeological records of them. Vegans have to take care not to injure their children's growing brains with a lack of protein by making sure they eat enough beans. Ultimately, veganism is an expression of deep empathy for other animals.

Vegans also don't wear products made of leather. You can buy vegan hiking boots that are made of plant fibers. Cats, as you know, are carnivores. What does a vegan do if they have a pet cat? Feed it vegan cat food of course. It is now commercially available. Feeling sorry for a cat is low on my list of priorities in life, but that almost does it. As an aside: if I were to suddenly shrink to six inches in height, my pet dog would try to protect me from my pet cat that would be trying to eat me. Which brings up another point. Some vegans have pet cats and dogs; yet, I have never seen a bumper sticker that says, "You can't call yourself an environmentalist if you have pet cats or dogs." In theory, just as with eating meat, many millions of malnourished people could be fed if those pet food resources were converted into grain for people. That is not really true, as I have pointed out in other essays. People go hungry because they can't afford food. Growing more does not help them. The granaries of the world are *presently* filled with food. The prices for some grains are already so low that our government must subsidize farmers to keep them in business. The answer to hunger is poverty reduction, not veganism. Veganism is a personal lifestyle choice, based on empathy and requiring discipline. One of my daughter's best friends was raised to be Vegan by her Vegan parents. They are a wonderful family; the world could use more people like them.

Veganism is at one end of the spectrum; on the other end we have the Atkins diet. Some Atkins fanatics profess that our ancestors were carnivores. Our teeth are evidence to the contrary. So is our digestive tract. If you want to see what the teeth of an omnivore look like, go look in a mirror. If you want to see the teeth of a carnivore, look at your cat or dog, a herbivore, look in a horse's mouth. In addition, the very fact that most people lose weight when they eat only meat and foods free of carbohydrates is suggestive that we are not designed to *thrive* on such a regimen. A diet that instructs people to eat nothing but grass would also have a very good record for weight loss because we are not designed to thrive on grass.

Most of human history has been a struggle to dodge starvation. There is no question that there have been subgroups of humanity who have lived primarily on animal protein—the Eskimos and Inuit for example. The ice age pushed other groups to live primarily on meat as well. Some Neanderthal groups probably subsisted for the most part on megafauna. We have been using fire to cook our food for hundreds of thousands of years, and this has allowed our omnivorous diet to include just about anything. Fat reserves are crucial during a time of famine.

We are designed to eat sweet foods and foods high in carbohydrates specifically to build fat reserves. America's struggle with obesity is the result of our food-worshipping culture—eating too much and exercising very little—not so much the kinds of food we eat. Witness the cable TV channels and magazines that are dedicated solely to food, the high percentage of food related commercials and the ever-growing number of culinary schools for specialized food fixers—chefs. Excuse me for digressing so.

In conclusion, the environmental movement has a tendency to be diluted by its affiliation with unrelated elements like animal rights and veganism. Will we soon see a disharmonious split between the environmentalists who eat meat and the animal rights proponents? Has that split already occurred? This propensity to splinter into sub-groups is part of our human nature. ELF, ALF, PETA, NPG, ZPG, secular humanists, agnostics, atheists, and brights, as well as the over 2,400 sects of Christianity are just a few examples of this phenomena. Human history is one long string of struggles between competing groups of people.

As a side note, a similar intermingling is happening to the Republican Party and the religious right. Over time, the two have become inextricably intertwined. A vote for the Republican Party is now a vote for the religious right. The Republican Party has become dependent upon the votes of the religious right to remain viable. The distinction between the two is quickly blurring. Like a one way ratchet, the Republican Party has been forced to become more and more dependent on the religious minions voting as their televangelist leaders tell them to. A takeover is underway. Combine that with unbridled indoctrination in home schools and America is in danger of being ruled by religious ideology rather than reason. Our constitution's philosophy of the separation of church and state has held this takeover at bay until now but appears to be losing ground under our present administration.

City ecosystems

Seattle has gone to great lengths to enhance and protect wetlands inside its borders. I am familiar with several of these projects. Many millions of tax dollars have been invested. Unfortunately, some of these areas are, with the exception of visiting waterfowl, sterile, and almost devoid of wildlife. Being completely isolated by highways and development, they are tiny islands and extinction events happen inside them very quickly. Wildlife cannot migrate to them or away from them. Thirty one million dollars have been spent since 1999 to improve the quality of Seattle creeks in an attempt to bring salmon spawns back. Think about it. That is a hell of a lot of money. The problem is that something is quickly killing the salmon after they enter these creeks.[150]

Many of us would like to live in a city, surrounded by wildlife and salmon runs, but is that realistic? Isn't that attempting to have your cake and eat it too? Look at the millions of cars lining the streets. Each car drips oil, antifreeze, grease, brake fluid, and window-cleaning fluid. Each one has four tires that are being worn down. Yards are full of weed killers, slug death, insecticides, and fertilizers. Look at all of the houses that line the streets. Each has paint that breaks down over time. Each house has a roof. Almost all roofs are covered in tar based shingles, each one degrading in the sun and washing down the gutters. All of this goes from the street, to the nearest stream or lake as surface runoff. It was standard practice in the last century to dump one's used motor oil into one's yard in areas where weeds needed to be controlled. Some people still do it. I wonder what could be killing the salmon? Cities are for people. Animals that can thrive inside the city are already thriving there. They don't need to be introduced. Attempts to reintroduce species to please our sensibilities will most often fail and will always be expensive. If Salmon could exist in a given steam they would already be there. Some streams still do have salmon runs.

Seattle's ecosystem is rich by some standards. On any given day, I can sit on my balcony and count at least twenty species of birds—woodpeckers, cormorants, herons, jays, crows, gulls, sparrows, wrens, robins, pigeons, geese, ducks, eagles, hawks and kites. Sitting on my vacation property in the middle of the Tahuya forest, I could probably count four or five species. In Seattle, I routinely see raccoons, rabbits, possums, and gray squirrels. There are also plenty of rats and mice. The waterways

---

[150] http://seattlepi.nwsource.com/local/107460_coho06.shtml

have beavers, muskrats, painted turtles, fish, and crayfish. The amount of wildlife per square mile in Seattle is easily five times what I would find in the woods at my country property. This high density is the result of the energy thrown off by human activity in the form of garbage, composting, gardens, ornamental plants and fruit trees. In addition, buildings provide a huge surface area for nesting and dens. The largest predators in the city ecosystem are house cats. There isn't as much diversity in a city ecosystem, that's for sure, but there is a high population density. Many, if not most of these animals are introduced species, but that fact is true everywhere you turn in today's mixed up world. Cities are ecosystems for people. Maybe we should accept the animals that have the capacity to live among us instead of trying to force other species to do so. Maybe we should be spending these millions of dollars to preserve intact ecosystems instead of opening up creeks buried under mall parking lots to make attractive but ecologically sterile memorials to what used to be. Spend it buying up or protecting habitat on the outskirts of a city, not in the middle of it.

Seattle is also home to a place called Discovery Park. It was once a military base. That is how it escaped development. This 534-acre chunk of land is bordered on one side by Puget Sound and on the other sides by Seattle neighborhoods. In addition to the ocean beach, it has two small ponds. It also shares space with Seattle's sewage waste treatment plant.

It is used in the summertime for children's day camps. It is also heavily used by joggers and dog walkers at all times of the year and is crisscrossed with dozens of hiking trails and several roads. Although leashes are required, and poop scoop laws are in effect, I once counted eighteen piles of dog crap in a half-hour walk. I know… I need to get a life.

One of the centerpieces of this park is a newly constructed nature center. It is the backbone for the children's day camps and staffed by a combination of childcare providers and budding young naturalists. Some see this park as an urban wildlife preserve. Although useful for children's educational purposes and joggers, it is actually just an overgrown city park. Calling it a wildlife preserve is stretching the definition.

The park has a pair of nesting bald eagles as its centerpiece. Bald eagles are common in the Puget Sound area. They can be found in almost all of Seattle's parks that have big trees near water. Green Lake Park and Seward Park have nests as well. Discovery Park also has a small remnant population of mountain beavers. These are large burrowing rodents native to the Northwest. Then there is the typical urban entourage of moles, squirrels, possums, raccoons and domesticated rabbits—pets that have been set free in the park. It has no turtles, lizards, toads or salamanders and only a small population of emerald tree frogs. It is also missing all higher level predators—other than hundreds of neighborhood cats, some garter snakes, and a few birds of prey. About twenty years ago a cougar actually found its way to Discovery Park. It was immediately shot of course. An active railroad skirts the park along the ocean beach. Railroad right-of-ways provide a thin green belt that runs for hundreds of miles. That is probably how the cougar found his way into a park in the middle of Seattle. It is also an excellent example of what can happen if ecosystems can be connected together with uninterrupted green belts or corridors.

The naturalists occasionally talk of eliminating all non-native plant life in the park, which strikes me as a waste of effort considering that most of the plant life is non-native and the park is surrounded by more plants that will continue to seed themselves. They also talk of eradicating the domestic rabbit population. This park isn't a natural ecosystem by any stretch of the imagination. It's a little better off than New York's Central Park, but not by much.

Americans traveling from coast to coast have occasionally transported small vertebrates either intentionally or accidentally in their trunks or luggage for many decades and will continue to do so. The pet trade is also alive and well. There is no way to prevent it and it is too late anyway. The huge mass of humanity has thrown most ecosystems into a giant blender. It is naïve to think that the blender can be shut off, especially with a U.S. population heading for 300,000,000.

Some lakes in Washington have painted turtles. Most lakes, however, do not. I suspect that this is because they are not native to the area. I have seen thriving populations of painted turtles in small remote lakes surrounded by hundreds of square miles of desert. These turtles are most likely the descendants of the hatchlings that used to be sold in pet stores or through the mail all across the United States in the '50s and '60s. Legislation (that was the precursor to modern animal rights) made it illegal to sell turtles with shells less than four inches in diameter because of the high mortality rate associated with keeping them as pets.

Hawaii exposed

The Hawaiian islands have been under assault by man for about 1700 years. Some of the first colonizers brought coconuts, pigs, dogs, and rodents with them. Visitors to Hawaii arrive with preconceived fantasies about lush rain forests and tropical waterfalls. As long as you don't look too closely, that is pretty much what you will see.

Following is a more realistic description. I realize that there is a fine line between pessimism and realism. I will try not to cross it.

I recently took a trip to Maui. My family and I drove to a trailhead that led to a snorkeling beach. We parked our brightly colored rental car next to the others. The trail had been well worn over the years by millions of tourists like us. We passed through a lush tropical forest. Some of the trees had strange hanging branches reminiscent of a weeping willow. One tourist, while taking copious pictures, commented on how beautiful it was. "Yes, just lovely," I replied. The trees were dead and the hanging branches were actually the aerial roots of a parasitic climbing cactus introduced from Mexico. The lush green foliage was nothing more than the cactus plant using the skeleton of the tree it killed to lift itself high into the canopy for sunlight. The trees that had not been killed by the cactus had been killed by another species of vine. This shady, green jungle consisted entirely of dead trees smothered in two kinds of non-native parasitic climbing plants. It was an ecosystem in its death throes. Not wanting to pop this tourist's bubble, I kept these observations to myself. What will this place look like ten years from now when the dead trees have all fallen over?

We also encountered several feral cats along the way. Like all feral cats, they were hungry, and breeding. Feral cats are a serious problem for any ecosystem because they sustain themselves by eating anything they can catch. Why were these cats allowed to run wild here? Why hadn't the department of whatever come out and shot them? Well, for one thing, shooting feral cats is likely to bring the wrath of animal rights activists down on your bureaucracy. You would have to come up with a more creative and expensive way to remove them.[151]

The trail eventually led to an abandoned boat ramp and a cove with seventy or eighty tourists sitting about on rocks or snorkeling. I did a little exploration off the beaten path. What I found was a lot of toilet paper and flies—the inevitable result of thousands of tourists and no facilities.

First-time visitors to the Hawaiian Islands always get an earful of local urban legends. For instance, did you know that the shells from pistachio nuts would puncture a car tire? Do an experiment sometime. You'll find that your tires will smash them to bits—no problem. Here is another one: mongooses were introduced to the islands to control rats, but rats are nocturnal and mongooses are not. You're supposed to deduce that bringing mongooses to Hawaii failed to have any impact on the rat populations. Nothing could be further from the truth. Mongooses are closely related to weasels and ferrets. They're all designed to go down prairie dog, rabbit, or rat tunnels to get their quarry, usually when they are sleeping. They don't wait for their meals to saunter by. Mongooses have made a huge impact on rat populations. Rats have adapted to the presence of the mongooses by nesting in trees thus taking advantage of the mongoose's poor climbing abilities. Unfortunately, along with millions of cats, mongooses have also made a huge impact on native bird populations. Rats can be a horrendous problem in the tropics. The aluminum bands seen on the trunks of palm trees in most Hawaiian resorts are to keep rats from nesting in the tops of the trees. The rats are there, but the mongooses have done a very effective job of keeping their numbers in check. Rats also have the ability to become primarily nocturnal or primarily diurnal

---

[151] http://www.cnn.com/EARTH/9802/23/cats.birth.control/

depending on local predation patterns. Mongooses have forced most rats in Hawaii to nest in trees and live nocturnal lives.

One of the biggest environmental disasters involving the introduction of non-native species occurred on the island of Guam. Brown tree snakes were accidentally brought to the island as stowaways on aircraft. In the absence of any natural predators they multiplied prolifically and have decimated the local bird life, driving several species to extinction. They are so numerous that they often cause power outages by crawling into and shorting out power transformers.

Authorities in Hawaii are concerned that the same thing might happen there. Special snake sniffing dogs are used to search aircraft arriving from areas known to have brown tree snakes. I am certain that Hawaii has been subjected to the same numbers of stowaway snakes, but they have never managed to establish a breeding population. The ecosystems of the Hawaiian Islands are quite different from those of Guam. In addition, snakes are a popular item on the mongoose menu. We have all heard the story of Ricky-Ticky-Tavi, the mongoose who saved a family from a malevolent cobra. The brown tree snake might have a difficult time getting a toehold on Hawaii even without additional efforts to prevent it.

We spent some time on the wet side of Maui, staying several days in a rented house that faced the ocean. A feral cat hung out near our back door begging for handouts. On one occasion, I witnessed it catching a mouse. We also saw several rats in a nearby tree. The cat did not seem particularly interested in them. One night, my youngest daughter caught a giant cane toad. These toads are also an introduced species and their skin exudes poison. Up to 50 dogs a year die in Hawaii after chewing on cane toads.[152] We also found a Jackson's chameleon. Apparently, they are becoming very common. Some estimates claim that up to twenty new plant or animal species are introduced to Hawaii every year. There are many thousands of feral pigs running about as well as deer. Hawaii is a lost cause as far as protecting what is left of its native flora and fauna. Fences to keep out feral pigs and pollinating rare flowers by hand can only slow the inevitable. Native plants and animals that do not successfully find a

[152] http://www.amonline.net.au/factsheets/canetoad.htm

niche in the ever-changing habitat of the Hawaiian Islands are doomed.

Sustainable living

People who care, people who want to do something for the impoverished and for the wildlife of the planet have been groping for a solution for some time now.

Enter, sustainable living. The concept behind sustainable living is to choose lifestyles and technologies that will allow us to live in equilibrium with the planet's ecosystems—using low volume toilets, composting, recycling and the like. It sounds too good to be true and it is. It is not in our nature to behave in the necessary manner. People and free market forces are driven by a quest for profit, not sustainability. There are already too many of us. People will continue to dump garbage down logging roads simply because it will save forty dollars in dump fees.

Tied in with this concept are the ecological footprint tests that can be found on the Internet. Your ecological footprint is a measure of how much damage your lifestyle does to the planet. It relies principally on the amount of resources you and your progeny consume. These tests ask you questions like; How far do you commute? How much meat do you eat? How big is your house or apartment and how many people live with you? Your ecological footprint is calculated based on these answers. You can then hide in shame or rub it in depending on your score. The idea is that if we can all just get our ecological footprints small enough, the people of the world will be able to live in harmony with our natural environment.

As you might have already guessed, I see some flaws. For example, try to explain why China and India are such ecological disasters. China has over 50,000 kilometers of major rivers. 80 percent of these rivers are essentially dead—they are so polluted that they contain no fish.[153] Yet, the people in China and India have an average ecological footprint that is many times lower than yours or mine. The idea that we can fix our problems by lowering our life styles to match that of Chinese or Indian peasants does not appeal to me. Nor would it appeal to those admin-

---

[153] http://www.ips.org/Critical/Enviroment/Environ/env1209006.htm

istering the tests as they sit comfortably in a roomy, warm, timber framed dwelling in front of their computers. That is not what the test creators envisioned—a life of mind numbing subsistence farming, no Internet, and no hot showers. China's and India's destroyed ecosystems are the perfect experiments showing that lowering one's ecological footprint all the way down to that of a rice eating peasant will not save the planet. You cannot blame the state of their environment on American consumption either. To put it another way: what we eat, drive, and buy—computers, SUVs, and Internet services—has a secondary impact compared to sheer numbers of people.[154] Some have argued that China's and India's problems are the result of poverty. My response—define poverty. Ten people living in a one room hut, no hot water, and no central heat is one definition—but such a lifestyle would also win most footprint calculation pissing matches.

Guilt-free vacations

A few years ago, I received a letter from one of my relatives. He had just returned from a trip to Guatemala where he and his children helped build a house for a poor family. Included were photographs of his children alongside the family with the new cinder block house. The Guatemalan family consisted of the father, pregnant mother, and their four children—ages two through five. It is the latest fad I call the "guilt-free vacation." You spend four thousand dollars and burn up three or four hundred gallons of aviation fuel to fly the family to an exotic, tropical third world country to help build a house, instead of simply going to Hawaii. The family could have stayed home and donated that money to a food bank and volunteered to work at a nearby shelter for homeless families, but where's the fun in that? In fact, the thousands of dollars spent on the vacation could have been given directly to the poor family who could have used it to employ local tradesmen to build their house. It is very unlikely that his family's labor was worth the money they spent on the vacation. Will this poor family have fewer children because they now own a cinder-block house? I think it unlikely, although I am

[154] http://www.nature.com/nsu/030721/030721-11.html

aware that some would beg to differ. I'm curious about his future plans. Will he fly his family back to Guatemala and build a house for each of the five children? Had his project included funding for family planning of some form his contribution would have been more meaningful. Another guilty conscious assuaged and another road to hell paved.

Environmental ethics of new immigrants

Environmentalism can be low on the list of priorities for some first-generation immigrants. They often come from parts of the world where the environmental degradation is many times more advanced than what is found here. They simply haven't had time to absorb our environmental ethics. Immigrants comprise the vast majority of the brush-picking workforce in Washington. Most of the litter found along logging roads is from brush pickers. The damning evidence is the type of litter left behind. Typically it is mixed with the telltale signs of brush picking like the rubber bands that are used to tie up the picked greens.

I knew an engineer at Boeing who was an immigrant from Vietnam. He embraced America's work ethic with a vengeance. Although he was a full time engineer by day, at night he worked as a police officer. This left him five or six hours a day to sleep, which seemed fine with him. One day he showed me several Polaroid photos from his relatives who had settled in the southern part of the U.S. along the Mississippi. The pictures were of pickup trucks filled with full-grown alligator snapping turtles with their heads chopped off. I was aghast. These turtles must have weighed over a hundred pounds each and were probably over half a century old. This turtle is considered a threatened species in most states. I asked him how they managed to catch so many. He told me that they used to trap turtles in the Mekong with woven baskets submerged near the shore. Turtles seek the baskets out because they resemble hollow logs. They had simply substituted metal barrels for baskets.

European style

I recently watched two wildlife documentaries produced in Spain. One dealt with wolves and the other dealt with bears. It turns out that the few remaining wolves subsist almost entirely on garbage. The documentary makers were surprised to find out that the wolves made regular nightly visits to cities in order to eat garbage. Without human garbage, these wolves wouldn't even exist.

The bear documentary was just as depressing. The only bear they could find to film was one with three legs, the fourth leg having been lost to an illegal trap.

Many people have the mistaken view that humanity's destruction of the planet is a relatively new phenomenon. Actually, we have been steadily destroying it for tens of thousands of years. The ancient Romans used to capture lions from the surrounding countryside for use in their gladiator games. Much of Italy was once covered with forests of spruce. The wildlife is long gone and the forests were all converted into wooden war ships. The microenvironment caused in part by the lush forests changed, leaving Italy and the Greek islands with a much more arid climate. This pattern continues unabated today and may eventually be the fate of the entire planet.

CHAPTER 7

Intelligent primates and culture

The third upper tooth is elongated and more pointed than the rest and is called the canine. Prominent display of their teeth is an instinctive behavior used to indicate non-aggression. Males are larger and typically more aggressive than the females. The thick mane of hair on their faces is another easily recognized feature unique to the males. They have excellent binocular vision. The nails on the feet and hands are thin and flattened. The hands are well adapted for grasping. They have no tails. They spend the majority of their time on the ground, but can climb quite well. Females typically give birth to one offspring at a time, but multiple offspring are not rare. Lactation occurs just after birth and may continue for three or four years. They have an omnivorous diet. They live in social groups. They are not on the endangered species list.

The fact that these primates display their teeth as an act of non-aggression should have been the hint you needed since most mammals show their teeth for the opposite reason. Maybe showing our modest canines to each other—smiling—has become a way of saying "I couldn't hurt you with these if I tried." On the other hand, what sequence of fight or flight responses run through a pet rabbit's nervous system when a child bares all of her teeth as she bends down to pick it up?

When given this riddle, very few people pick Homo sapiens as the primate described because most of us see ourselves as separate from the animal kingdom. We are primates. Most primates have nails instead of claws at the ends of their hands instead of paws. All have a set of canine teeth and two eyes that face forward for enhanced binocular vision. Most have vocal chords used to make a variety of sounds. We don't have tails and neither do the other great apes.

Each primate species has unique differences that set it apart from the others: gorillas are the largest, howler monkeys the loudest, mandrills the most colorful, another the smallest and on and on. There have been thousands of primate species which are now extinct. Some were giants standing over nine feet tall. Many of those species walked upright just like us. At different times in the past, there have been upright-walking primate species co-existing on the planet, paying little attention to one another much like chimps and baboons ignore each other today. The famous Lucy skeleton is one example. Picture a skinny chimp with long slender legs standing around just like a person. Fossilized foot-prints of these upright-walking primates have been found on two separate occasions. We are descendents of one of those species. It just so happens that today we are the last remaining primate species that walks upright. We share the distinction of walking on two of our four appendages with the birds.

Many anatomical features are used to differentiate primate species from one another: existence of a tail, hair patterns, hand and foot structure, size and shape of teeth. Our only anatomical feature that warrants real interest is the large size of our brains. Just as an elephant's long nose and large size sets it apart from the other land mammals, we are set apart by our bulbous heads. That's pretty much it. A human brain by itself is not very impressive. The complex cultures that have evolved as a result of these big brains however are more than impressive; they have become the destroyers of nature.

Humankind smugly looks at the computer screen in front of him, at the airliners flying above him, and revels in his great intellect. However, those examples of technology are the result of huge, unbelievably complex cultures, not an increasing intellect. We have big heads, literally and figuratively. The Neanderthals had bigger brains than we do. One cannot correlate smartness solely to brain size, but the evidence does not increase the odds that we have been getting smarter. Our present day super-cultures got their start because of a combination of population size and density, which facilitated trade with other large groups. This provided an economic incentive for people to specialize and a technology explosion started ten thousand years ago that has never stopped.[155] We certainly didn't suddenly get smarter.

I recall a beer commercial where an actor, portraying an aerospace engineer, pats the nose of a business jet he had just finished engineering. Yes sir, time for a well-deserved beer. In reality, it took decades of accumulated knowledge and thousands of people to design and build just the black plastic nose cone on that jet. The idea that one human being could design an entire jet is hilarious to any experienced engineer—engineers being easily entertained.

Look at one of the buttons on your shirt. I challenge any human being on the face of this planet to duplicate from scratch that simple little button without any help from another person. Before rushing to take up that challenge, consider the following. The button is probably made out of plastic; a polymer created from petroleum pumped from the bowels of the Earth. The button was formed in a mold that was, in turn, machined from a metal alloy. The machine tools used to make the mold were made of several different metal alloys. How are you as an individual going to mine, smelt, mix and purify metal alloys and then use them to build drilling jigs with carbide or diamond tipped drills to extract the petroleum? How will you make the machine tools used to fabricate the mold used to make that button? How would you manage to turn the raw petroleum into plastic? Truth be told, duplicating from scratch a simple plastic button is a task that is orders of magnitude beyond anything any human being is even remotely capable of doing *alone*. It takes the combined knowledge of several generations of humans, and tens of thousands of individuals working in concert to make a plastic shirt button.

Termites cannot survive alone and neither can we. Drop a single termite worker alone into any field and it will perish in very short order unless it can find its way back to its colony. The same is true if you drop a person into the middle of any unfamiliar wilderness. Survival would depend on how long it took to find other people who still had a link with their culture. As individuals, we really are not all that bright.

---

[155] Nonzero by Robert Wright

Procreation is the machine that drives evolution. Those who use their intellect to deliberately limit the size of their families will limit their genetic future. In other words, if people with lower IQ scores are more successful at having babies, then we should eventually see a drop in IQ scores. Most people view this idea as a negative. There is no reason to do so. Traits like empathy, generosity, enthusiasm, and creativity are not linked to IQ.

However, IQ scores have been going up dramatically since their invention—and this can in no way be accounted for by evolutionary selective pressure in just two generations. James R. Flynn is a professor of political science at the University of Otago in Dunedin, New Zealand. He is responsible for unearthing the Flynn effect.[156] He found that a person who scores average on an IQ test today would score as a genius if he took a test from the 40's. Something does not make sense and that something is the test itself. We have not all become geniuses. My opinion is that whatever IQ tests are measuring, they do a poor job of documenting the capacity of the human mind—wisdom, intuition, empathy, creativity, motivation, and awareness. All domesticated animals—from dogs to cows—are much less intelligent and have smaller brains than their wild counterparts.[157] There is no question that domestication would have the same effect on Homo sapiens and may have already. Our wild ancestors may have been brighter in many ways than we are today or will be in the future.

Eugenics is a euphemism for animal husbandry applied to people. Selective breeding of domestic animals isn't a science. It generates more failures than successes and requires the culling of mistakes. That alone makes the idea of applying it to people completely unacceptable. Eugenics is the attempt to domesticate human beings. Secondly, one man's improvement is another's degradation. Imagine trying to get a consensus on what traits constitute the perfect dog. There is no such thing. A sheep dog is good at herding sheep and a rat terrier is good at catching rats. A hybrid of the two wouldn't be good at either task. In addition, the

---

[156] http://www.sciam.com/article.cfm?articleID=00037F65-D9C0-1C6A-84A9809EC588EF21

[157]http://groups.msn.com/alanboylescosmiclog/112002archive.msnw?action=get_message&mview=0&ID_Message=304

genetic problems associated with different breeds would fill a large textbook. Eugenics has nothing to do with evolution or survival of the fittest. Blue-eyed, blond-haired people are getting rarer every day. Ironically, we have Nazism to thank for killing the eugenics movement. It didn't die because people suddenly realized how irrational the idea is, it died because the Nazis made it very politically incorrect to promote it or be associated with it. Eugenics was based on a deep ignorance of both genetics and evolutionary selective pressure. That ignorance still exists. Evolution was used as an analogy for social engineering, and is still a smoldering ember. Type it into your favorite Internet search engine and prepare to be amazed. If you do so, you will also find creationists trying to associate the theory of evolution with Hitler's master-race scheme.

I have access to data that Darwin did not. I know for example that evolution does not proceed in a linear direction. It reverses, changes course, goes forward again, and sometimes stops completely.[158] It also does not proceed toward perfection. The animals alive today are not more complex or advanced over the dinosaurs of sixty million years ago. They are just different. The circulatory and immune systems of the dinosaurs were probably just as complex as those of modern animals. Birds are not descended from the dinosaurs—they are dinosaurs. Darwin did not know those things and neither did the proponents of eugenics. The very fact that schizophrenia is so prevalent suggests that it is imparting—at least the milder cases of it—some reproductive advantage or it would have been weeded from the gene pool long ago. Maybe most shamans were schizophrenics and had large numbers of wives? We would be wise to leave well enough alone. Einstein would have been culled as a youngster and probably a lot of other Nobel Prize winners as well. If evolution is putting the breaks on human intelligence—whatever that is—would it be a bad thing? We are on the verge of destroying the planet's biodiversity with our technology now. Nature will take its course. We are not nearly as bright as we think we are. We most certainly are not bright enough to domesticate ourselves.

Our genetic programming to have sex, compete for higher status, and make war on members of other primate groups is the

---

[158] See <u>The Beak of the Finch</u> by Jonathan Weiner

same as many other social primate species. It is only that our complex cultures and resulting technologies make us appear different. We tend to slaughter the opposing members of a primate group by the millions when we go to war, but that is only a matter of magnitude. We have overpopulated our environment just like any other very successful animal. This is further evidence that we are not separate from the rest of the animal kingdom; we just think we are. The cultures we surround ourselves with are all that set us apart.

What is intelligence? The dictionary definition of intelligence speaks of the speed at which one can learn, of the ability to learn and understand, problem solving skills, responding successfully to new situations. It is a woefully inadequate definition of overall intellect. The weird, socially awkward, but really smart egghead is a well-known stereotype. We've all heard stories of autistic people with amazing mathematical or musical abilities. These unfortunate people demonstrate examples of hyper-developed facets of intellect. Your local egghead has the same problem, but not as exaggerated—a hyper-development of some aspect or aspects of intellect at the expense of others. Is genius a result of brain damage? Mozart heard notes as colors. I suspect that Einstein was probably not so bright in many respects—his musical ability, his ability to empathize, to communicate non-mathematical ideas, to navigate through social nuance and subtlety, to make art, to remember names and faces. More than likely, he would have been diagnosed with Asperger Syndrome today and kicked out of the public school system as a child. I am only making an educated guess, not having known the man, but if he followed the general pattern for genius, he was probably just as dumb in many respects as he was brilliant in others. My apologies to you, Albert, if I have offended. The news media of his time propelled him into celebrity status. It is still very common to find his image in magazine ads selling just about everything, insinuating that the smart buyer will buy a particular this or that. You can readily buy a poster of him to hang on your wall if you want to gain some measure of status by association. Be careful, some of us might mistake your intent and assume you have Asperger syndrome.

CHAPTER 8

Battle of the sexes

Evolution has been waging a battle with big-brained primates to keep us from planning our pregnancies for hundreds of thousands of years. Once we grew smart enough to connect the act of sex with pregnancy, the battle was engaged. At some time in our evolutionary past females probably came into heat and displayed this fact in some manner, as most other mammals do. Once hominid females figured out that mating while in heat could create a pregnancy, they learned to avoid sex at that time if they wanted to avoid a pregnancy. The same strategy held true for males who wanted to avoid making babies.

Evolution has created a whole bag of tricks to counter our ability to dodge pregnancies. One is the simple fact that human females come into heat very discreetly, providing no clues for those who want to avoid sex on days of maximum fertility.

Another trick is female orgasm. While orgasm is used throughout nature to motivate males into having sex, it is not an option for most female mammals. Human females however, are by far the most likely to experience them. This exception is probably the result of evolutionary pressure countering our ability to avoid pregnancy. By increasing sexual receptivity among females with this added enticement, there would be a corresponding increase in the number of unplanned pregnancies.

There are strong arguments that this unique ability of the human female to climax is just a residual system, like male nipples, and that there is no evolutionary selective pressure for it to exist. The fact that most women cannot consistently climax through natural intercourse is considered part of the evidence supporting the idea that female climaxes are an anachronism—they are just a potential.[159] Although this was most likely true in ey are just a potential.[159] Although this was most likely true in our evolutionary past, it is no longer so.

A woman has to experience this phenomenon only once in a while to realize that sex holds a potential for something very pleasurable. This unique ability provides additional incentive to motivate the only primate smart enough to connect the act of sex with conception to participate in that act.

Why is it so much easier for men to have them? Again, evolution has created this discrepancy to maximize conception. It is imperative for conception that the male climaxes first and that the female remains receptive long enough for this to happen. If women climaxed first, their sexual satiation would cause a tendency to break off intercourse before insemination. Typically, the female hangs in there seeking a climax—which may never arrive—guaranteeing that the male is satiated. Fairness is not a necessary quality of evolutionary selective pressure.

You might think that if women were as motivated to have sex as men you would get a lot more unplanned pregnancies—and indeed you would—too many in fact. Evolutionary selective pressure is much more complicated than just having babies. These babies must be successfully reared to have children of their own. Females have to be less motivated to have sex than males, or they would always end up with more babies than they could care for.

To put it yet another way, men live in the pre-climactic state. They enter the post-climactic state only for a short time after having sex. Women live for the most part in the post-climactic state, entering the pre-climactic one much less often. Women are the sexual gatekeepers—the moderators of sexual behavior. Men are generally ready to have sex when the opportunity presents itself. It isn't a matter of maturity. It is a matter of genetic programming. It is the genetic role of females to put the brakes on sex.

Look at what can happen when the brakes are removed. Some male homosexuals have hundreds or even thousands of sexual partners in a single year. So, what's wrong with that you say? Isn't that what every heterosexual man would want if

---

[159] Female potential All about Eve by Elizabeth Loyd, http://philsci-archive.pitt.edu/archive/00000853/00/chapter_2.pdf

women would just cooperate? You have a point there, but transmission of disease is the problem. AIDS is primarily a heterosexual disease. It is prevalent in gay men because of two factors. First, without women in the equation, their male propensity to have so much sex with so many partners has no braking action. Second, gay men have understandably concentrated in big cities such as New York and San Francisco, which greatly facilitates this propensity. Dense population concentrations of homosexual men have very high rates of sexually transmitted diseases—hepatitis, gonorrhea, herpes, syphilis—you name it.

Disease may be the root of male homosexual taboos in many cultures. As with many taboos, a physical reason often exists for it somewhere. Taboos are usually embodied in secular laws or religious doctrine at some point. Without modern medical technology, a dense population of gay males in a place like ancient Rome would have quickly burned itself out with sexually transmitted diseases. What I have said here will be taken by some to be anti-gay. It isn't. Homosexuality is a genetic fact—it is a natural variation, like hair color. Huge concentrations of gays in giant cities are, however, unnatural. Homosexuality has been preserved by evolutionary selective pressure. It is imparting some kind of reproductive advantage. In a tolerant, free society, especially a modern one with modern medical technology and education, it should be given freedom of expression in the name of individual happiness.

Why do women who live together end up with synchronized menstrual cycles? The answer has something to do with reproductive success, that much you can bet on. Exactly what the benefit is, no one knows. Men tend to have multiple wives if given the opportunity. This propensity may be linked in some manner to menstrual synchronization. Like many other social primates, human males tend to be polygamous—men have a subtle urge to collect females. Most modern cultures, however, do not usually allow expression of this drive.

Does pre-menstrual syndrome (PMS) exist for a reason? How could this contribute to procreation? That's a good one. I will venture a guess or two. Does a hyperirritable female drive the male out of her personal space leaving her vulnerable to sexual advances from another male—causing her to have one or two more children than she would otherwise have had? It is surprising how many children are not related to their fathers. I've seen figures suggesting that up to 20 percent of impoverished children in the U.S. have been fathered by males outside of the marriage.

Statistics have uncovered other subtle tricks used by Mother Nature to drop pregnancies into our laps. For example, many of them occur at the beginning or end of relationships. Now, why would that be?

It is a common phenomenon for a man who is about to lose his mate to another, to suddenly find his partner more sexually desirable. This behavior undoubtedly helps to explain the higher than average number of unplanned pregnancies that occur toward the end of relationships. Could this be nature's way of planting a seed before it's too late? Do women sometimes become pregnant in a subconscious attempt to keep a mate?

Genes control the imprinting instinct seen in geese and ducks. Oppositely, familiarity causes sexual indifference in siblings, and unrelated children raised together in a Jewish Kibbutz. This suggests that a gene for controlling incest may switch on or off.[160] I suspect that a similar phenomenon occurs in all human sexual relationships. This theory has the potential to get me in very hot water with my wife of eighteen years, so I must be very careful here. Most marriages in our Western culture end in divorce. Those that remain intact are not the passionate liaisons they started out as—contrary to what magazines at the checkout stand want you to believe.

The honeymoon phase of any sexual relationship is very short lived. Why is that? What causes an initial sexual attraction to wane? Familiarity is certainly a necessary part of the equation. Again, whatever the reason, it has to do with reproductive success. If people were instinctively monogamous, the honeymoon phase would never end. They would look across a crowded swimming beach, spot their spouse and think "Oh yeah, that's what I like." It happens, but not all of the time. In the primate world, gibbons are the most monogamous. The males and females are the same size and color and have similar social behaviors. Gorillas are on the other end of the spectrum and are

---

[160] Nature VIA Nurture by Matt Ridley, page 171

polygamous. The males are much larger than the females, more aggressive, and less sociable. It is obvious that human behavior falls somewhere between these two extremes. Without social restrictions, men clearly have a tendency for polygamy. Men are also larger, less sociable, and more aggressive than females, but these differences between males and females are not as extreme as those seen in gorillas. It is interesting to me that nature comes up with so many different reproductive schemes.

Jealousy is a powerful and overwhelming instinct. This aggressive instinct forces us to be possessive toward a mate, but there is no equivalent instinct that makes us have eyes for only one mate. A built-in tension somehow works to maximize reproductive success in people, but causes friction within marriages. An analogy would be to put two dominant male gorillas in the same zoo enclosure with a harem of females. Two mature, unrelated, silver-backed males trapped in close proximity goes against the grain of gorilla programming. This would be a big mistake, and every zookeeper knows it.

A new term was coined to describe our reproductive pattern—serial monogamy. Genetically programmed instincts attract us to a mate and bind us to them through a combination of affection and jealousy for an extended period, after which other instinctive urges and tensions cause a tendency to break off one relationship for another. What are some of the physical mechanisms that put this process in motion?

Anyone who has lost a lover knows how painful such an experience can be. Depression is often part of the aftermath. Could fear of depression based on experience help to coercively bind some relationships together? Is this the reason depression has not been weeded from our gene pool? Maybe PMS plays a role in the process to break couples apart, again, suggesting a reason for it to remain in our genetic makeup.

Virtually all self-help books dealing with marriage say the same thing—maintaining a marriage takes a lot of effort. Depending on the personalities involved, a harmonious marriage usually requires a conscious commitment by both parties to remain married, and to find ways to be happy in that marriage by hook or crook regardless of biological programming that pushes us to do otherwise. I'm in the doghouse now. People who break off a marriage because of waning sexual attraction caused by familiarity are embarking on a hopeless quest. It doesn't matter who you choose for your next mate, your sexual passion for that individual will quickly level off. That is not to say that married couples can't have good sex lives together, but lust is not the word to describe it. Happiness is something you must strive for inside and outside of a marriage, regardless of evolutionary driven predilections that attempt to lead us about by our noses. Consciously recognizing those predilections goes a long way toward overriding them.

The feminist movement in the United States successfully provided women with greatly improved economic and career options in our complex industrial culture. One tactic taken to change these social mores was to play down the differences between men and women. That particular strategy however, has its limitations and may become a barrier to further progress. Acknowledging and understanding the differences between men and women may hold the key to further advances in male-female parity in our hopelessly complex societies.

Men and women have many obvious physiological differences. These differences however are just the tip of the iceberg. The main schism between men and women is in their wiring. Most women have no idea what goes on inside the head of a typical man. They would probably be appalled. The same can be said for men understanding women. The following comments are from the American Journal of Gender Studies:

> How can you tell if a man is aroused?
> He's breathing.
>
> What is the insensitive area at the base of the penis called?
> A man.
>
> What do men and women have in common?
> They both distrust men.
>
> Women dream of world peace, a safe environment, and eliminating hunger.

> Men dream of being stuck in an elevator with the Doublemint twins.
>
> A woman knows all about her children. She knows about dentist appointments, soccer games, romances, best friends, favorite foods, secret fears, hopes, and dreams. A man is vaguely aware of some short people living in the house.

I wasn't serious about these comments being from a fictional journal (I hope it was fictional). These are "men" jokes taken from the Internet, and I am sure that some will find them offensive.[161] I have chosen not to counterbalance the aforementioned jokes about men with jokes about women simply because I have probably offended enough people already.

This leads to the subject of stereotyping. We subconsciously play the odds every day of our lives. It is an instinctive and necessary part of our nature. The only way you can avoid stereotyping is to carry a lucky coin to flip every time you have to make a decision. Anyone who did that would quickly find his or her life degenerating into a train wreck.

Let's assume that you want to buy a used car. You find two identical cars in the paper. A little old lady is selling one of them. She claims that she only drove it to church and back, that it has low miles, and has been kept in the garage its entire life. A used car salesman who says the same thing is selling the other. Do you get your lucky coin out? The word may have a negative connotation but the behavior called *stereotyping* is instinctive and is not going to go away. That used car salesman may have been telling the truth but unless you have plenty of time to spare, you will start your search by looking at the car owned by the little old lady. You have stereotyped used car salesmen.

Now, let's assume that you run a child daycare center. Right next door is a huge pool of cheap labor, a men's penitentiary. Your city government has a desperate need to employ these guys as they get released. You have been told repeatedly that stereotyping is bad. There is a college for women across town. Your

---

[161] http://www.deeplake.com/men/jokes/men_vs_women.shtml

instincts tell you that you should hire from the college. You have now stereotyped the inmates. Do you cave in and hire ex-cons?

Stereotyping has its dark side of course. If you have a strong racist tendency and the college is exclusively for African American women, you might make the mistake of hiring only white parolees and live to regret it. Grouping individuals by race instead of by criminal record or gender was the fatal flaw made by our racist daycare center entrepreneur. Like the old programmer saying goes: garbage in, garbage out.

The other problem is what to do about that one convict who would make a wonderful employee but will not get the chance because you don't have the resources to find him. It's an imperfect world. Looking for the grain of wheat in all that chaff might bankrupt you. Just being conscious of this instinctive predilection to group individuals will help you to spot the exceptions when they cross your path. When the politically correct tell us not to stereotype, what they should be saying is that you should always keep your eyes open for the exceptions, that statistics apply to groups, not individuals, and that you had better have your facts strait before you stereotype. But that's a mouthful. It’s a lot easier to say, "you should not stereotype." Playing the odds is an innate and necessary survival instinct. We all do it, everyday of our lives as part of our human nature.

On average, women have superior social skills—name and face recognition for example. Manual dexterity in the form of fine motor skills is another well-documented difference. Statistically speaking, women are more empathetic, nurturing, generous, and loving towards people that they know. Women are just better designed for certain things, like being a caregiver or mother. Babies find the warm, smooth, soft skin of a woman very comforting. This is a result of mutual programming. Most women are perfectly capable of piloting an Apache attack helicopter and most men are capable of being a gentle caregiver for a baby—sans breast-feeding. Statistically speaking, however, the majority of men would find it more difficult to care for that baby than a woman would. In many ways, women are just smarter than men. As a side note, some gene must have switched on as I have gotten older. I find babies, toddlers, and young children absolutely irresistible. May I live long enough to become a grandfather.

Most social mammals exhibit obvious physiological and behavioral differences between males and females. Male lions have a mane and are larger than females. A male lion's job is primarily to fight with other male lions and to mate with the females in his pride. He will also kill any cub that was not sired by him. Male lions are bad mothers—literally and figuratively. Evolution has created this difference in male/female lion behavior just as it has created differences in male/female human behavior. Men, however, are not lions and are not necessarily bad mothers.

The world would look very different without men behaving like men—different, but not necessarily better. Men start and fight almost all wars, commit almost all violent crimes, and build almost everything you see. Just about every tree cut down and wild animal shot or trapped is at the hand of a human male. Look out your window. You will see houses, cars, roads, skyscrapers, and airplanes. They were almost all conceived, designed, and built by men. Some might suggest that these objects are the result of male domination. Male domination is very real, primarily because men are statistically more aggressive. Formalizing this domination with fundamentalist religion or secular laws exacerbates the problem, but male aggressiveness alone cannot account for our proclivity to build stuff.

Why are most women less interested in building structures and machinery? Only an idiot would suggest that women are incapable of such things. My hypothesis is that men in our culture build things to attract women. Like wasps building a nest, they just can't help themselves.

Men do it to compete with other males for higher status. Status is primarily how men attract women. The space shuttle and the atomic bomb are all constructs resulting ultimately from competition among males. Competition for breeding rights is a common thread all through nature. Those things men build are ultimately monuments to women. Men (and women) are largely clueless about this fact. Women also compete and seek status of course, but they are not quite as compelled to do so as men or in the same manner. Women are not men with different shaped bodies. That simple fact that our bodies *are* so different is ample evidence that we also have different behavioral propensities.

Most men don't hesitate to ogle a member of the opposite sex that they find attractive. Women rarely bother. The most important job in human culture is the care and nurturing of our children. It is a difficult, time-consuming endeavor. This job falls almost universally on the shoulders of women simply because, statistically speaking, they are better at it than men. Women could and would build things if they were so motivated, but for the most part, they just aren't. Although completely capable of doing so, *statistically speaking*, they simply do not have the genetic incentive to create these monuments; it's not how they are made.[162] Keep in mind that there are a number of women's engineering societies, all staffed by exceptional women engineers. Statistics apply to groups, not individuals.

These behaviors are generalizations, and there are many, many exceptions. Again, a prosperous and generous culture will make room for these numerous deviations in the name of individual freedom and happiness. That is what women's liberation is about, and it is a good thing for all of us. The differences we see between men and women have all been grossly magnified by our gigantically complex mega-cultures. Hunter-gatherer societies don't have the means to grossly exaggerate these sexually driven dimorphisms, but the differences still exist. San bushmen work hard at being good hunters not so much because they need the food, but so they can impress the women back in the village and hopefully parlay their status as a good hunter into some hanky panky—men, go figure.[163]

---

[162] http://www.massnews.com/past_issues/other/4_Apr/aprmt3.htm

[163] See Human Evolutionary Psychology, by Louise Barrett, Robin Dunbar, and JohnLycett

CHAPTER 9

Programming and status

I have concluded of late that we are less creatures of choice than we are creatures of design. Why do we allow sixteen thousand children world wide to die of malnutrition related illnesses every day? Programming. We are programmed to care most for those who are closely related to us. Caring for people of other cultures on the other side of the planet will always be a low priority from an evolutionary, genetic point of view. Like balancing one bowling ball on top of another, natural predilections can be overridden with conscious intellectual effort, but such efforts cannot be sustained in perpetuity. In other words, we can be convinced to occasionally cut a check for a relief agency, but we will not write very many of them. If parents have the choice of spending $19,000 a year to send their child to a private school, or donating that money to a non-profit organization for starving children in Africa, all will opt for the private school.

"Hold on there just a darn minute," you say in a fit of guilt. "One-half of one percent of the tax dollars I'm forced to pay goes to foreign aid!"[164]

I say to you, big whoop. You'll need to come up with a better argument than that to convince me.

The same line of reasoning holds true for our treatment of other species. There are roughly 219,000 humans born every single day. This explains why there are only 400 Siberian tigers and a few hundred California condors left in the wild. The list of animals driven to extinction and soon to be driven to extinction by man is long and growing rapidly. There isn't anything intrinsically evil about our behavior; it's perfectly natural. There are simply too many of us.

---

[164] http://www.usaid.gov/about_usaid/

The seeking of higher status within a primate group is typical behavior for many social primates. An alpha male and an alpha female have the best chance for successfully raising their young. This behavior can reach truly perverted proportions in our modern day mega-cultures. Witness the monuments past and present day people have built for themselves or what every wealthy individual stereotypically builds for themselves in the form of outrageous homes. Most Americans will go deeply into debt to secure the largest home and most impressive car within their means. This is ample evidence of this subtle and underrated instinct in action. It fuels our whole economy.

This inborn desire to seek higher status is as strong and real as our instinct to bare our teeth when we want to show non-aggression—smile. It is as strong as our drive to have sex or eat, yet it goes largely unrecognized. It motivates our day to day behavior more than any other drive. Like sex, status is rarely talked about in polite company, yet like sex, it motivates almost everything we do.

Status takes many forms. Groups of airplane pilots put the captain of a 747 at the top of their status hierarchy. The Vatican, doctors, lawyers, engineers, and school children all have their social hierarchies. None of us are immune to it.

I know of one group that puts great status in replacing commonly accepted symbols with their own, thus deluding themselves into thinking that they are immune to this instinctive drive. Members of this group can be identified by bumper stickers on older model cars that say things like "Kill your television" and "Visualize World Peace."

You increase your status whenever someone envies you for something you have, can do, or have accomplished. It's as simple as that.

One interesting thing about status is that it's always changing, and outside of your peer group, your status symbols become largely meaningless. An airline captain may be at the top of his hierarchy only to find that, because he does not drive a VW van with a dead head sticker on the bumper, he impresses no one at the hemp festival he accidentally stumbled into. All healthy, normal people continuously seek higher status. A person without this drive is not mentally healthy at all. Monks who sequester themselves in caves as hermits devoid of all material wealth are an odd bunch, but they have found a way of achieving higher status in this act (you can also bet that monastery infighting can get pretty hairy). Recognizing status-seeking behavior for what it is can go a long way toward understanding why we do the things we do—write books for example or build monuments to ourselves in the form of opulent homes.

Few people in our culture admit to seeking status. For one thing, it is an instinctive urge. As such, it tends to manifest itself at a subconscious level. Most of us are not even aware of our behavior. If you ask the owner of a luxury SUV why he didn't buy something just as functional, but much less expensive, you won't hear this for an answer: "I bought it because I felt it would make many of my peers envious. It is also a symbol of my ability to obtain expensive luxury items, which is in turn an indicator of my overall prowess." No, you are more likely to hear, "I needed the power, can't live without lumbar support, and the all-leather interior is also nice."

It is considered bad sportsmanship to openly admit to status seeking. You can't admit that you are deliberately trying to impress people with your ability to make money, art, write a book or whatever. That would be like showing your hand at the start of a poker game—just not smart. From an evolutionary perspective, we seek status because it gives us pleasure. The pleasure is caused by chemicals in the brain. The chemicals are a reward for participating in an activity that may attract a mate, which may in turn make a baby. Of course, we rarely do things *solely* to obtain status. The SUV is also a form of transportation, writing a book is also a way of conveying ideas. But the car you drive, the landscaping in your yard, the art you create, the remodel you did, are all forms of advertisement telling passersby of your abilities.

For many people, men in particular, parties serve as a backdrop for competition. The first thing many people do at a party is check out the competition and try to establish where they fit in the pecking order.

"What do you do for a living?"

"I'm the head of neurosurgery at children's hospital. How about yourself?"

"I sell paint."

Of course, the paint salesman may be very good at what he does and may make twice as much money as the surgeon. He'll have to find a way to leak that data if he wants to improve his standing. To flat-out tell the surgeon how much you make risks having him tell you he makes twice as much. In our culture, one's income is the single biggest measure of status. Your home and the objects in it are also sources of status. A view has high status. Remodels are done primarily to increase one's status. A home that is paid for has higher status than one with a new mortgage. Status permeates everything we do. You cannot ignore it and remain happy and healthy. We are genetically locked into competition. It consumes most of our energy.

The definition of what has high status is slippery and ever-changing. The old saying "keeping up with the Joneses" implies that the competition has no end, and it doesn't. You can't rest on your laurels.

Like a poker game, status is also rife with deception. Did he pay cash for that luxury SUV or does he live in it to afford the payments?

I recently asked a retired friend of mine what he thought about the whole status-seeking thing. He sagely told me that he doesn't get caught up in all that crap. We were riding in his brand new Cadillac SUV at the time, one of two that he owned, heading for his two million-dollar condominium with a view of Puget sound.

You can earn status, stumble on it, have it handed to you, inherit it, and even get it by proxy. Status is no longer an accurate measurement of ability in our modern complex cultures. However, it still attracts mates. In a monkey troop, the high status males and females have the highest reproductive success. I doubt that correlation exists any longer in our culture. It has become an anachronism. In other words, status seeking may have lost much of its meaning from an evolutionary point of view. High status couples do not have better reproductive success in America. It doesn't matter; we are still driven to seek it. Those with lower status seeking urges may now have an edge on those who practically kill themselves for it. If this trend were to hold long enough, it would affect the gene pool and move evolution in a new direction. Would the world be a better place populated with people who have a low status-seeking urge? Maybe Mother Nature knows what she is doing.

Status is a form of competition. You will seek the level you are comfortable with—like basketball players who gravitate to ball courts where they can compete. You won't seek out a court where you dominate or will be dominated. You will find your peers.

You might say to yourself, "I don't want to compete. I'm dropping out of the game." That might be a big mistake. Look at what often happens to the high-powered executive who retires. One minute he's king of the hill, the next, he's just another old guy in the grocery store, and a very unhappy one at that.

Hormone levels in the brain depend on perceived status. Your mental and physical health depends on it. Males who fail to compete can have miserable lives, as can males who compete and fail. It's a catch-22 and we can't escape it. It is part of our genetic make up.

It is a fact that females shape the behavior and even the looks of males over time by their selection criteria. The male peacock is usually held up as an example of this phenomenon. Any cock that can haul a load of feathers like that around and still survive in a jungle full of leopards and tigers is one tough bird. Women are not immune to status seeking and competition, not by a long shot. When a woman selects a mate she does so, in part, by subconsciously evaluating how he will be perceived or envied by her peers. The same can be said about men. Trophy wives they call them.

A woman's ability to attract and pick a mate is a form of competition. A handsome fry cook will usually lose out to the homely rich guy.[165] What if that handsome fry cook also played the guitar and knew every Beatles song by heart? What if he had a warm, empathetic, and generous personality, and was on the board of directors for the society against child abuse? What if the rich guy had inherited his money and was narcissistic and incompetent? The woman might be happier with the fry cook but may never know him well enough to learn about his strengths. Her initial attraction is primarily through status. Now, like I said

---

[165] See article on status anxiety, June 16th 2003 issue of Newsweek

before, status is slippery and ever changing. A good-looking man carries a lot of status in some circles, as does intelligence and creativity.

Very few of us realize that our basic nature is leading us about by our noses. We think we have a lot more free will in some areas than we actually do. When we open our minds, push away our defense mechanisms, and start asking why we do what we do, the answers are revealing. There is only one reason teenage boys learn to play a guitar: guitars are chick magnets. Why do we invest so much energy and time maintaining our lawns? Is it because we want them to look nice? Why do we want them to look nice? As you stand there with a beer, looking at your freshly mowed lawn, you jump into your neighbor's head and envision his envy. Yeah baby!

Medical researchers do not go to work everyday to benefit humankind; they do so primarily in a quest for status and a paycheck.

Untold numbers of species that exist today will be driven to extinction in our children's and grandchildren's lifetimes because of this subtle piece of primate programming called status seeking.

CHAPTER 10

A state of mind

One might ask what a discussion of depression has to do with overpopulation, human nature, or the preservation of biodiversity. I am going to use it as an example supporting the concept that we are motivated by chemical nuance at least as much as we are driven by rational thought. This ties in with human nature and why we have grown indifferent to the concerns of a world losing its biodiversity. The late Julian Simon—mentioned in an earlier essay—suffered chronic depression for thirteen years. His depression lifted when his work began to bring him recognition and acclaim. He eventually wrote a book describing how he believed he had cured himself.[166] I have not read it. I wonder if his optimistic point of view was his way of defeating his depression. Few thoughts are more depressing than those that foresee the end of nature. The thoughts and ideas in this essay are my own, based on my single experience with a depression during my adolescence.

I believe that depression sufferers see the world at its most basic level—as it really exists. Depression sufferers may actually have a clearer view of reality than the rest of us and this clear and accurate view is why they are so unhappy. The rest of us are seeing the world through rose colored glasses. I do not mean to say that depressed people have the proper, natural, or correct view, just one without important filters. An analogy: a welder has to put goggles on to avoid damaging his eyes. Does that mean he has the correct view of the flame? No, he has a view of it that will not cause him harm. A welder cannot safely view a torch flame without his goggles because his eyes are not capable

---

166 Good Mood: The New Psychology of Overcoming Depression by Julian Simon

of dealing with such intensity. It was suggested to me once at a dinner party that only a pessimist would take this point of view. People who have never endured a true depression have no idea how devastating it is. They think of it as a severe case of pessimism. They also tend to describe pessimistic people as being depressed. Pessimism and depression look similar on the outside, but they are not related. If you have never been truly depressed you are probably incapable of comprehending just how horrible it is. The fact that it so often ends in suicide should be ample evidence supporting my position.

I would like to propose a hypothesis and suggest that all of us view life through a chemical filter, or a series of daily natural highs. The drugs responsible for these natural highs are chemicals found in the brain. I've seen heroin addicts, teeth rotted out, abscesses in their skin, HIV positive, high as a kite, happily discussing their situation with some investigative reporter. Happy, because they are viewing reality through an *artificially induced* drug haze. We sit in front of our television sets aghast that these poor souls can be so content considering their horrible reality. Now picture a depression sufferer looking through the living room window, watching you watch your television, aghast that you can be so content considering *your* horrible reality, which unfortunately, is also his reality. Healthy people have ways of dealing with unending bills to pay, yards to maintain, stressful commutes to boring jobs, inevitable aging, illness and death in a world being stripped bare of all biodiversity by teeming masses of other human beings. A depressed person does not.

Can an individual be forced into a state of depression through exposure to the appropriate physical environment? In other words, if one took an optimistic happy individual and put him into a long term, hopelessly miserable environment, could a state of depression be temporarily induced? Can our defense filters protect us only to a point? Were the victims of Nazi concentration camps and gas chambers generally a depressed group of people? Is depression a common problem for people who have just been told that they have a very short time to live?

Of course some people are more susceptible to depression than others. A poorly understood condition called postpartum depression sometimes follows childbirth.[167] How long a person remains in a state of depression is highly variable. The situation required to induce a state of depression would also vary from individual to individual. The latest pop psychology mantra claims that people are either born happy or not and their environment has nothing to do with it. There is some truth to this. Evolution works at many levels. The huge diversity of human personalities is an example of this. There are many niches to be filled in human cultures. Serious or moody people have their roles and rewards as do the exuberant bubbly people who bounce about doing their thing.

There is a wealth of information being accumulated on the effects of brain chemistry and drug addiction. Research is suggesting that our moods are primarily, if not totally, the result of chemicals in the brain. Put another way, happiness, sadness, anger, elation, and a general sense of well being are all the result of chemicals being dumped into our blood stream. What are these chemicals, and what do they do for us?

Serotonin is one. It is associated with high status and competitiveness. Your levels of serotonin are higher when you are king of the hill. It makes you feel good about yourself. It is the chemical primarily responsible for corporate ladder climbing behavior—status seeking.

Endorphins are naturally occurring mood enhancers and pain killers that come on-line—make you high—in times of great physical stress just prior to death, or as you go into shock after a severe physical or emotional trauma. Nobody feels pain when they are in shock. A person in shock has been injected with high levels of endorphins. Chemical releases in the brain are also what make ordinary activities like sex and eating enjoyable.

To a depressed person, eating is the act of stuffing a tasteless nutrient paste into a hole in his face. Now, extrapolate this eating analogy to all normal daily activities and you have life as a depressed person experiences it. Human beings are, by design, just a little bit high all the time. These daily pleasures are naturally induced by chemicals as a stimulus or reward for doing things that—statistically speaking—keep us alive and procreating. That high varies daily with our activities.

---

[167] http://www.postpartum.net/

People with bipolar disorders—persons suffering from manic and then depressive mood swings—give us some insight. The pleasure center of a manic-depressive is malfunctioning, putting out too much happy juice one-day and not enough on others. I have a friend who used to be manic-depressive. I have repeatedly observed him through the years go from a depressed mood to a happy exuberant one and back again in the span of a few days. As he has aged his highs and lows have moderated greatly. His mood now sits in the normal range for the most part with only an occasional excursion into mild depression or mania.

Can one be too happy and optimistic? Overly optimistic people can get themselves into a world of trouble. My friend once decided during a manic episode to fly from San Francisco to Louisville, Kentucky to retrieve his aging, senile father from a nursing home. On the plane ride back my friend's mood passed through the normal range and he found himself sitting next to an eighty-year-old demented man who didn't even recognize him. Having very limited financial means he was forced to fly his father back to Louisville a few days later. Make enough bad decisions like that and you will find that your life is a train wreck—the rarely mentioned down side to pop psychology's mantra "the power of positive thinking." If positive, upbeat personalities were superior to others then the world would be populated with nothing but positive upbeat people. Evolution obviously has other designs. Overly optimistic people were often the last ones to see the lion coming.

Why don't we treat depression sufferers with doses of cocaine or heroin to make them happy again? The answer to that question is addiction. It is suspected that mood-enhancing drugs neutralize our natural pleasure centers. That is why it is so difficult to kick a drug addiction. Drug addicts find that when they try to quit, life no longer has any pleasure. They are essentially depressed. It is as though their chronic use of mood enhancing drugs has caused their natural pleasure centers to atrophy through lack of use.

The intense pleasures associated with sexual intercourse are rewards for engaging in activities that promote the duplication of your genetic material—making babies. The pleasure we get from eating sweets is a chemical reward for seeking out and eating a naturally rare and very efficient source of calories and energy. Honey for example, is a universally sought after and treasured food in all hunter-gatherer societies. Primates have their own special emotional needs as compared to, let us say, cats. The human primate has even greater needs. As the intellect of Homo sapiens developed so did the need to protect that intellect from the harshness of the world. Our ancestors had to have mechanisms in place to alleviate the anxiety of potentially being eaten by a predator during the night. Our self-defense chemicals and behaviors have developed in tandem with our intellect.

We have developed many very elaborate means of filtering reality. Coping mechanisms we call them, that bag of tricks we use to get by from day to day—prayer, denial, and laughter to name a few. Most of these weapons are missing from the depressed person's arsenal of self-defense. Many of us need help dealing with the anxiety generated by the realization that we're going to die someday, and that it will be very unpleasant. We are the only animals on the planet with enough computing power in our heads to foresee these events.

There are those who think that the pervasiveness of religion may be explained as a specific cultural behavior that releases anxiety-retarding chemicals in the brain. Religion of one kind or another is pervasive in virtually every human culture. In evolutionary terms, people willing and capable of suspending rational thought to believe in a given religious dogma may have some advantages that allow for more efficient procreation—bringing children into the world and successfully raising them to have families of their own. I personally believe that it is more complicated than that. There is no such thing as a religion gene. A propensity to subjugate oneself to a strong leader who is on your side, maybe. Formalizing that tendency with organized religion is probably a perversion of what the original behavior evolved for.

The correct way to look at a depressed person may be that he is seeing the world for what it really is, while the rest of us are seeing a sugar coated version of it. A human being cannot survive for long looking at pure uncut reality, just as a welder will not see for long without proper goggles. People fear death. People fear the idea of suicide and often try to stigmatize it by calling it a coward's way out. Cowardice has nothing to do with it. Anyone capable of overcoming his or her own self-preservation instincts is suffering from a serious malady. Depression is a uniformly fatal affliction if it isn't rectified in time—unique in that the mechanism of death is always suicide.

CHAPTER 11

The Future of Life—a review

While doing research for my book, I consistently ran across references to Edward Wilson's *The Future of Life*. I thought it would be a good idea to read it after I finished writing mine, just to see if I missed anything important. I purchased it a few days after I finished Volume I.

Wilson has written a wonderful book. Make no mistake about it. I highly recommend it to anyone interested in preserving the planet's biodiversity. He covers a depressing topic yet leaves one with a sense of hope. After reading it, I turned to my wife and said, "Wilson might think I read his book before I wrote mine."

I could hardly believe my eyes when I came across his discussion of poison dart frogs on page 122. He mentioned a frog named Phyllobates horribilis. What I think he meant to say was Phyllobates terribilis. Horribilis is the last part of the Latin name for grizzly bears. It is reassuring to know that even a person of Wilson's stature (and his editors) can make mistakes. I suspect it was corrected in later versions. By the time I finished reading his book, I had found over a dozen other references that our books shared. I have since made a few alterations based on Wilson's book now that I have read it. My guess is that since the books are both on the same topic—preservation of the planet's biodiversity—there are bound to be similarities.[168]

Volume I of my book mentions the word overpopulation just twice—the same as Wilson's. His book is eloquent and conciliatory in tone. I suspect that he deliberately avoided using the word

[168] I have since read several books after having written this one and I have found that they all contain many similarities to my book. I should not have been so concerned.

overpopulation in the name of diplomacy—I certainly did. Most conservation groups have accepted the idea that we must simply try to preserve as much as possible regardless of population growth. They no longer have an official position on population. The word overpopulation has a negative connotation—it leaves a bad taste in the mouth. Our books are quite different in most respects. Aside from the obvious fact that volume I of my book is fictional, we do not see entirely eye-to-eye on the solution to the problem. Continuing our conservation efforts is an obvious part of any solution, and one that we both agree on. We differ in that he is hoping for the spontaneous rise of a global environmental ethic:

1) "In order to pass through the bottleneck a global ethic is needed." Page XXIII
2) "... we may be ready to settle down before we wreck the planet." Page 22
3) "Let us make an honorable assumption that economist and ecologist have as a common goal the preservation of life..." Page 24
4) "...environmentalism ... is not yet a world view... (like) sports, politics, religion..." Page 40
5) "... a universal environmental ethic... (is necessary). Page 41."

I cannot envision environmentalism ever becoming as popular as sports, politics, and religion unless a way can be found to make it meet the same basic human desires—competition, status, and reassurance in the face of impending mortality. A global ethic of environmentalism with enough influence to do what needs to be done is also unlikely to develop. I hope I am wrong on both counts and that Wilson's vision pans out. Hedging my bets, however, I think we must focus on finding more ways to channel human nature in directions that lead to a preservation of the biodiversity as a side effect. Eco-tourism and the TIFIC are prime examples of this concept.

A large part of Wilson's book is dedicated to building a powerful case that it is cost effective to preserve biodiversity at sustainable levels. Repeatedly he suggests that we must save nature so it can be put to use. I don't believe Wilson himself really cares so much that we save nature so we can consume it. He wants to save nature one way or another and is trying to build a rational case from an anthropocentric point of view. His arguments give strong reinforcement to those of us who want desperately to preserve what is left, but in my opinion, they are unlikely to bring about drastic change because rational argument has far less impact on the behavior of Homo sapiens than what you might expect.

Wilson's closing paragraph is as elegant as the rest of his book:

> "I hope I have justified the conviction, shared by many thoughtful people from all walks of life, that the problem can be solved. Adequate resources exist. Those who control them have many reasons to achieve that goal, not least their own security. In the end, however, success or failure will come down to an ethical decision, one on which those now living will be defined and judged for generations to come. I believe we will choose wisely. A civilization able to envision God and to embark on the colonization of space will surely find the way to save the integrity of this planet and the magnificent life it harbors."

Wilson praises "A civilization able to envision God and to embark on the colonization of space... " Like Galileo before him, Wilson wisely patronizes the religiously inclined majority. In addition, the billions spent by the military/space industrial complex would go a long way toward securing threatened ecosystems and maybe even propel a TIFIC into reality. Instead, it sends people and things at great expense and risk into orbit around a dying planet. We are not going to colonize the stars anytime soon.

Time is running out and nobody knows this better than Wilson. He has appealed to religionists and to conservative economists, but the problem is, they aren't going to read his book. They aren't going to read mine either. In addition, very few people share his near-religious love of nature. Even those of us who do, barely hesitate to write books that we know will consume a couple of trees (in my case) or thousands of them (in Wilson's case). We continue to have sex that may result in accidental pregnancies, drive our children back and forth to school, line our home's floors with oak, and build decks of cedar. We are human beings and we cannot always control through intellectual effort our instinctive predilections. I take a harder view of humanity. I believe that solutions have to be compatible with our instincts and our nature if they are to have any hope of long term success. As I have said before, the motives to *make* the TIFIC would be profit and status. The motive to *take* it would be to dodge unplanned pregnancies. The fact that it will help save the planet's biodiversity will be a side effect. Human beings are simply not programmed (it is not in our nature) to make significant personal sacrifices solely to save the planet's biodiversity or even to help less fortunate members of our own species. If we are going to pull it off, we will have to find ways to save it despite ourselves.

Index for Volume II

Visit www.PoisonDarts.net for up-to-date internet links to the footnotes and forum discussions of biodiversity preservation.

I also encourage you to join the discussion on ways to preserve our biodiversity by writing a review of this book. To do so, find Poison Darts on Amazon.com. and click the "review this book" hyperlink.

You can sell your used book back to Emerald City Resources LLC by following the instructions found on www.PoisonDarts.net.

Made in United States
Troutdale, OR
11/03/2026

56852524R00226